Springer Water

The book series Springer Water comprises a broad portfolio of multi- and interdisciplinary scientific books, aiming at researchers, students, and everyone interested in water-related science. The series includes peer-reviewed monographs, edited volumes, textbooks, and conference proceedings. Its volumes combine all kinds of water-related research areas, such as: the movement, distribution and quality of freshwater; water resources; the quality and pollution of water and its influence on health; the water industry including drinking water, wastewater, and desalination services and technologies; water history; as well as water management and the governmental, political, developmental, and ethical aspects of water.

Mariacristina Cocca · Veronica Ambrogi ·
Roberto Avolio · Rachele Castaldo ·
Maria Emanuela Errico · Gennaro Gentile
Editors

Proceedings of the 3rd International Conference on Microplastic Pollution in the Mediterranean Sea

Springer

Editors
Mariacristina Cocca
Institute of Polymers Composites and Biomaterials (CNR IPCB)
National Research Council of Italy
Pozzuoli, Italy

Veronica Ambrogi
Department of Chemical, Materials and Production Engineering (UNINA DICMAPI)
University of Naples Federico II
Naples, Italy

Roberto Avolio
Institute of Polymers Composites and Biomaterials (CNR IPCB)
National Research Council of Italy
Pozzuoli, Italy

Rachele Castaldo
Institute of Polymers Composites and Biomaterials (CNR IPCB)
National Research Council of Italy
Pozzuoli, Italy

Maria Emanuela Errico
Institute of Polymers Composites and Biomaterials (CNR IPCB)
National Research Council of Italy
Pozzuoli, Italy

Gennaro Gentile
Institute of Polymers Composites and Biomaterials (CNR IPCB)
National Research Council of Italy
Pozzuoli, Italy

ISSN 2364-6934 ISSN 2364-8198 (electronic)
Springer Water
ISBN 978-3-031-34454-1 ISBN 978-3-031-34455-8 (eBook)
https://doi.org/10.1007/978-3-031-34455-8

This Springer imprint is published by the registered company Springer Nature Switzerland AG
The registered company address is: Gewerbestrasse 11, 6330 Cham, Switzerland

Contents

Chapter 1
An Experimental Comparison Between FTIR and Raman Microspectroscopy Applied to the Morphological Analysis of Microplastics in Drinking Water

Luca Maurizi, Lucian Iordachescu, Inga V. Kirstein, Asbjørn H. Nielsen, and Jes Vollertsen

1.1 Introduction

One of the most widely used analytical techniques for microplastic (MP) chemical analysis and quantification is Fourier-Transform InfraRed micro-spectroscopy (μFTIR). In particular, those equipped with a Focal Plane Array (FPA) detector, can take extensive chemical images of plane surfaces in a relatively short time, even if complex analytes such as fibers are present [1].

Raman micro-spectroscopy (μRaman) is another spectroscopic technique, which is being increasingly used in the MP science field. It is able to obtain a far finer spatial resolution in the xy plane than μFTIR [2], therefore it is generally utilized for the smallest particle fraction down to about 0.4 μm, depending on the chosen analytical parameters. On the contrary, μFTIR is the preferred technique for particles above 5 μm [2]. μRaman-based methods normally employ a so-called "point and shoot" approach, where the instrument's laser is automatically driven onto each identified particle, which is distinguished from the visible image's background by contrast [3].

The present contribution aims to compare these two techniques by presenting preliminary MP morphological data from a field study conducted at a Danish drinking water plant.

L. Maurizi (✉) · L. Iordachescu · A. H. Nielsen · J. Vollertsen
The Department of The Built Environment, Aalborg University, Aalborg, Denmark
e-mail: lucam@build.aau.dk

I. V. Kirstein
Alfred-Wegener-Institute Helmholtz Centre for Polar and Marine Research, Biologische Anstalt Helgoland, Helgoland, Germany

M. Cocca et al. (eds.), *Proceedings of the 3rd International Conference on Microplastic Pollution in the Mediterranean Sea*, Springer Water,
https://doi.org/10.1007/978-3-031-34455-8_1

1.2 Experimental

1.2.1 Sampling

The investigated plant (anonymous) produces drinking water from groundwater by means of aeration and sand filtration with no further chemical treatment, given the low concentration of pollutants in the Danish groundwater sources. The groundwater source lies 92 m in depth, so a network of pumps and pipes transfers the water from the source to the distribution network.

1 μm pore size, sintered steel filters (Mesh Masters, The Netherlands) were employed to filter the drinking water at the facility. Filtering devices and holders were custom-made. The sampling devices consisted of four separate flow lines in stainless steel along which the filter holders which could be mounted in series. The inlets of the two devices were connected to the water network by means of a flexible steel pipe. The sampling took place from the 28th of September to the 1st of October 2021 (approximately from 9 am to 2 pm on each day) at both the plant's inlet and outlet. About 1 m^3 of drinking water was filtered for each field sample. The exact sampled volume was checked by means of the flowmeters mounted at the outlets of the two devices' flow lines. 30 field samples were obtained in total and subsequently analysed first with μRaman and then with μFTIR.

1.2.2 Methods

1.2.2.1 Preparation of Samples

The filters were conserved in clean Petri dishes in a flow bench until use. Each filter was incubated in 5% SDS (Sodium Dodecylsulphate, VWR, Germany) at 50 °C for 24 h in a thermal bath (Thermo Fisher Scientific, Germany) inside a clean beaker, then the particle–enriched SDS solution was filtered through the same filter in a glass filtering system connected to a vacuum pump. The glass components in direct contact with the filter were thoroughly rinsed with 50% ethanol (Ethanol for HPLC, Th. Geyer GmbH, Germany), then the filter was removed, put in a new beaker, and covered with a layer of 50% ethanol. The beaker was covered with aluminum foil to prevent external contamination and ultra-sonicated for 10 min.

A new glass filtering unit was equipped with the same filter and the particle–enriched 50% ethanol solution was filtered through it. The system's glass funnel was again rinsed with 50% ethanol, the filter removed, put in a clean beaker, covered with a layer of either clean >99.5% ethanol (μRaman samples) or 50% ethanol (μFTIR samples), and ultra-sonicated for 10 min to detach the particles from the filter.

Finally, the filter was scratched with a clean spatula and rinsed with a small amount of clean ethanol to manually recover the particles and transfer them into the ethanolic solution.

The particle-enriched ethanol solution was poured into a clean 10 mL glass vial. The vial was finally put in an evaporator and kept under gentle nitrogen flow in a water bath at 55 °C until exhaustion of the entire volume. The dried samples were kept in the fridge until use.

The μFTIR samples were reconstituted with 2.0 mL of clean 50% ethanol, of which 1000 μL were deposed. The deposition was performed with a glass pipette on 13 mm diameter ZnSe (Zinc Selenide) windows (Crystran Ltd., UK). The samples were dried at 55 °C on a heating plate overnight, which produced an active area of about 10 mm diameter, prior to the μFTIR analysis.

The μRaman samples were instead reconstituted with 1.0 mL of clean >99.5% ethanol. 25.0 μL were deposed in triplicate through a home-made metallic deposition funnel. The deposition was performed with a glass pipette on 10 × 10 mm Si (Silicon) substrates (SmartMembranes GmbH, Germany). The samples were dried at 55 °C on a heating plate overnight, which produced an active area of about 2 mm in diameter.

1.2.2.2 Analytical Techniques

The μFTIR analysis was conducted by means of a Focal Plane Array (FPA)-based Fourier Transformation Infrared Spectroscopy (FTIR). A Cary 620 FTIR microscope coupled with a Cary 670 IR spectroscope (Agilent Technologies, USA) was used to scan the entire area of the enriched ZnSe transmission windows. The microscope was equipped with a 25× Cassegrain objective producing 3.3 μm pixel resolution on a 128 × 128 mercury cadmium telluride (MCT) FPA detector. All scans were carried out in transmission mode with a spectral range of 3750–850 cm^{-1} at 8 cm^{-1} resolution applying 30 co-added scans in transmission mode. A background tile was collected before each sample's scan, using the same parameters but co-adding 120 scans instead of 30. A chemical image of the sample's active area was obtained, where each pixel contained an IR background corrected spectrum.

The μRaman analysis was performed in Dark Field with a Renishaw InVia (Renishaw plc UK, Wottonunder-Edge, UK) with the following analysis parameters: laser wavelength 532 nm, grating density 1800 ll/mm, montage objective 50× long-distance, analysis objective 100× long-distance, 50% laser power, 0.4 s acquisition time, 10 accumulations. The objects in the active area were distinguished according to the intensity threshold set at the beginning of the visible image analysis and filtered by setting "Length" and "Width" above 1 μm.

1.2.2.3 Data Post-processing

Chemical images from the μFTIR analysis were analysed with the software siMPle [4]. It can automatically detect the particles in a sample and extrapolate their morphological parameters. Each pixel containing an IR spectrum was compared with the software's library, which led to an estimation of Pearson coefficients for each comparison. Both the original spectrum and the first derivative were compared with the library's references. Moreover, spectral matches from the software were manually checked to further assess their reliability.

For the μRaman analysis, the experimental data were analysed with the machine's software. Renishaw's Polymer library was used for spectral recognition, while morphological data were obtained prior to the signal acquisition.

1.3 Results and Discussion

1.3.1 Results

The MPs identified with μFTIR were grouped into five length ranges (5–50 μm, 50–100 μm, 100–200 μm, 200–500 μm, and 500+ μm) and their mean frequency was calculated over the five investigated days for both the plant's inlet and outlet (Figs. 1.1 and 1.2). The MPs longer than 50 μm represented the most frequent items encountered in the samples, accounting for 85% in total. In particular, the 200–500 μm (32%) and the 500 + μm (51%) length ranges proved to be the most populated length fractions in the investigated days. Interestingly, the 5–50 μm length range accounted for 15% of the inlet samples MPs, whilst it was not represented anymore in the outlet samples. Indeed, in the latter case, the 500+ μm MPs dominated the distribution (74%), followed by the 200–500 μm length range (24%).

These findings are quite different from what was found by μRaman. Figure 1.3 graphically illustrates a comparison between the two employed techniques, by averaging the total number of MPs found at the plant's inlet and outlet. For readability

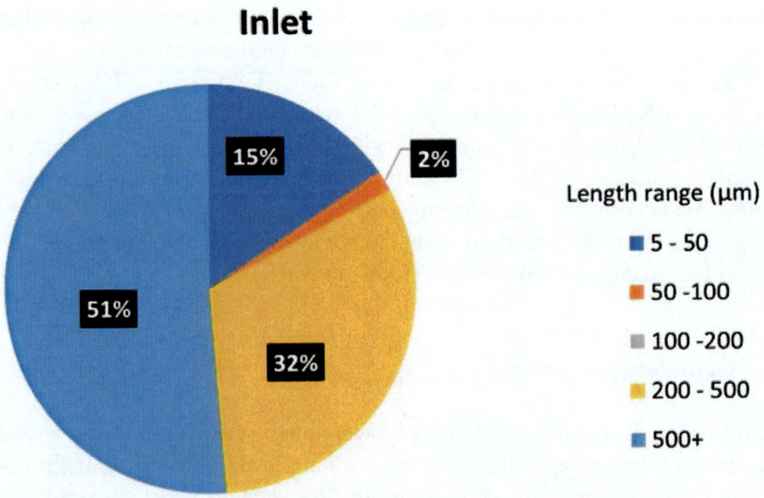

Fig. 1.1 MPs length ranges frequency over days 1–5 at the plant's inlet

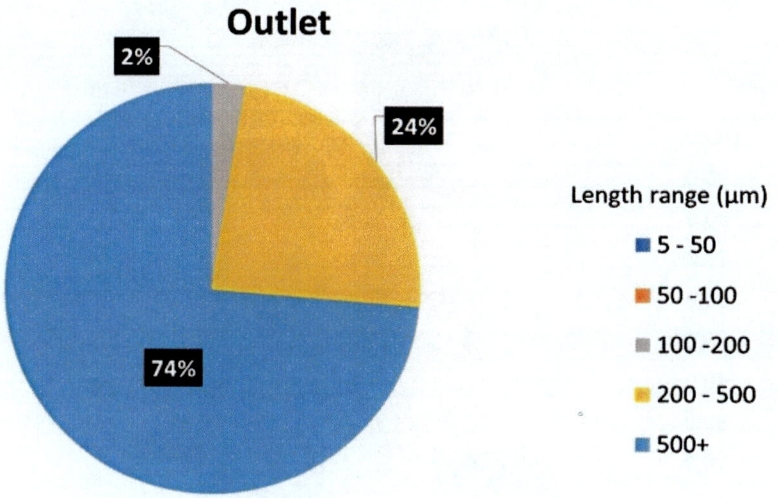

Fig. 1.2 MPs length ranges frequency over days 1–5 at the plant's outlet

purposes, the MPs were divided into two length ranges, below and above 50 μm. μRaman showed a dominance of MPs below 50 μm, which accounted for 99.8% of the total identified MPs. On the contrary, according to the μFTIR analysis, 88.2% of MPs belonged to the 50+ μm length range.

1.3.2 Discussion

μFTIR and μRaman proved to perform differently during the morphological analysis of the samples, with the μFTIR being scarcely able to locate and analyse the MPs below 50 μm. On the other hand, the μRaman dataset was characterized by a far lower number of MPs above 50 μm. The first reason for this finding can be the curser spatial resolution of the μRaman, which is usually addressed as the main advantage of the technique over μFTIR [5]. Therefore, the μRaman dataset is mostly populated by the smaller particles since they are likely the most abundant in the samples [6]. The second reason is given by the different sample deposition methods employed: for μRaman, 25 μL out of 1 mL were deposed (2.5% of the total volume), while for μFTIR 1000 μL out of 2000 μL (50% of the total volume). This means that the probability to depose larger particles was higher in the μFTIR protocol, given their scarce frequency in comparison with the smaller size fractions.

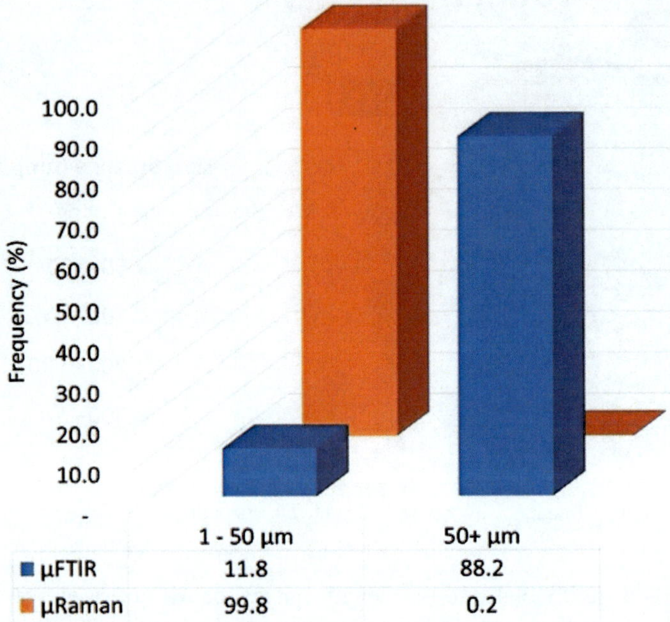

Fig. 1.3 Comparison between μFTIR and μRaman with regard to the MPs length data

1.4 Conclusions

The present contribution aimed to compare the analytical performances and the outcomes provided by the most widely employed spectroscopic techniques for the MPs analysis, the μFTIR and the μRaman. By comparing the data obtained from the two techniques, it was possible to assess that the <50 μm length range MPs were the most abundant in the μRaman dataset. On the contrary, the μFTIR could mostly identify MPs above 50 μm. Besides the results strictly relating to the investigated drinking water plant, we would like to stress the complementarity of the μFTIR and μRaman. As a further development of this work, the data from the two techniques will be compared in terms of MPs abundance over time and MP removal efficiency of the plant.

References

1. Primpke S, Dias PA, Gerdts G (2019) Automated identification and quantification of microfibres and microplastics. Anal Methods 11:2138–2147. https://doi.org/10.1039/C9AY00126C
2. Wang W, Wang J (2018) Investigation of microplastics in aquatic environments: an overview of the methods used, from field sampling to laboratory analysis. Trends Anal Chem 108:195–202. https://doi.org/10.1016/j.trac.2018.08.026
3. Araujo CF, Nolasco MM, Ribeiro AMP, Ribeiro-Claro PJA (2018) Identification of microplastics using Raman spectroscopy: latest developments and future prospects. Water Res 142:426–440. https://doi.org/10.1016/j.watres.2018.05.060
4. Primpke S, Cross RK, Mintenig SM, Simon M, Vianello A, Gerdts G, Vollertsen J (2020) EXPRESS: toward the systematic identification of microplastics in the environment: evaluation of a new independent software tool (siMPle) for spectroscopic analysis. Appl Spectrosc, Artic 0003702820917760
5. Ivleva NP (2021) Chemical analysis of microplastics and nanoplastics: challenges, advanced methods, and perspectives. Chem Rev 121(19):11886–11936. https://doi.org/10.1021/acs.chemrev.1c00178
6. Kooi M, Koelmans AA (2019) Simplifying microplastic via continuous probability distributions for size, shape, and density. Environ Sci Technol Lett 6:551−557. https://doi.org/10.1021/acs.estlett.9b00379

Chapter 2
Small Microplastics (5–100 μm), Additives, and Plasticizers in *Mytilus Galloprovincialis*, A Potential Bioindicator for Microplastics Pollution

Fabiana Corami, Beatrice Rosso, Davide Asnicar, Maria Gabriella Marin, Carmen Losasso, Andrea Gambaro, and Carlo Barbante

2.1 Introduction

As the definition reported by ECHA [1], microplastics (MPs) can hold within additives and plasticizers; they are part of microlitter together with natural and non-plastic synthetic fibers [2]. MPs were reported to be ingested by different organisms, from invertebrates to mammals [3–5]. Organisms ingest these particles according to their mouthparts' size; thus, invertebrates ingest MPs, additives, plasticizers, and other microlitter components less than 100 μm in size (i.e., SMPs and APFs). Through the invertebrates at the base of the trophic web, SMPs and APFs can reach other organisms, especially those at higher levels of the trophic web, with the chance of giving rise to bioaccumulation and biomagnification. Also, once entering the trophic web, these particles can pose a significant risk to human health. Coastal environments and transitional environments are areas of high-ecological diversity and substantial

F. Corami (✉) · B. Rosso · A. Gambaro · C. Barbante
Institute of Polar Sciences (CNR-ISP), National Research Council of Italy, Via Torino 155, 30172 Venezia-Mestre, VE, Italy
e-mail: fabiana.corami@cnr.it

Department of Environmental Sciences, Informatics, and Statistics (DAIS), Ca' Foscari University, Via Torino 155, 30172 Venezia-Mestre, VE, Italy

D. Asnicar · M. G. Marin
Department of Biology, University of Studies of Padua, Via Ugo Bassi 58 B, 35131 Padua, PD, Italy

C. Losasso
IZSVe, Italian Health Authority and Research Organization for Animal Health and Food Safety, Viale Dell'Università 10, 35020 Legnaro, PD, Italy

M. Cocca et al. (eds.), *Proceedings of the 3rd International Conference on Microplastic Pollution in the Mediterranean Sea*, Springer Water,
https://doi.org/10.1007/978-3-031-34455-8_2

productivity. Since these areas may also be a sink and source for MPs and other pollutants, they are among the critical habitats for monitoring, for which a sentinel organism, commonly found in large areas with biological traits that make it suitable for the purpose, can be employed. Mussels are usually considered optimal sentinels for this kind of study, being common in temperate coastal seas all around the globe, sessile, which can provide location-specific information and a link to the human food chain since they are very popular as seafood.

In this study, the preliminary results of the abundance of SMPs (5–100 μm), additives, and plasticizers in specimens of *Mytilus galloprovincialis* will be presented and discussed.

2.2 Experimental

2.2.1 Sampling Sites

Specimens of *Mytilus galloprovincialis* were collected in 19 different sites in the Venice Lagoon (Fig. 2.1). The sites are located in the three basins into which the Venice Lagoon is divided, as described in Table 2.1.

Specimens were collected from *bricole* or dolphins, which are nautical structures used to indicate waterways (canals) in the lagoons of Venice, also used for mooring vessels.

2.2.2 Methods

2.2.2.1 Preparation of Samples

Specimens were collected and stored at 4 °C until their arrival at the University of Padua laboratory. Once there, specimens were frozen and then transported to the Institute of Polar Sciences (CNR-ISP) laboratory at Venezia-Mestre, where they stayed frozen till the pseudo-digestion. Shells were measured with a digital caliper to assess their size class; then, the organisms were left thawed in a plastic-free clean room (Fig. 2.2), where the pseudo-digestion of organisms took place. The individuals were weighed and then put in decontaminated glass vessels. The pseudo-digestion was performed according to the pre-treatment method developed by Corami et al. [3]. The pseudo-digestion developed is a soft pre-treatment method to avoid further denaturation/degradation of plastic polymers, additives, and plasticizers. After the pseudo-digestion, the slurry was filtered on alumina oxide filters (Whatman ANODISC® (Anopore Inorganic Membrane), Whatman, Merck Darmstadt Germany, 0.2 μm, 47 mm) using a glass vacuum filtration system (VWR International, Milan, Italy) and a Laboport® pump. Filters were rinsed before the filtration using methanol,

Fig. 2.1 Map of the 19 sites where specimens of *M. galloprovincialis* were sampled

ethanol, and ultrapure water (Elga Lab water, High Wycombe UK). After filtering the slurry, filters were purified with isopropyl alcohol, methanol, and ethanol. Reagent and procedural blanks were filtered accordingly.

2.2.2.2 Analytical Techniques

Analyses were carried out using a Micro-FTIR Nicolet™ iN10™ (Thermofisher Thermo Fisher Scientific, Madison, WI, USA). By employing the software section PARTICLE WIZARDS, particles were analyzed, employing a liquid-nitrogen-cooled MCT detector (mercury cadmium telluride detector) in transmittance mode with a spectral range of 4000–1200 cm^{-1}, 100-μm step size scanning (spatial resolution), 100–100 μm aperture, and 64 co-added scans at a spectral resolution of 4 cm^{-1}.

Table 2.1 List of the 19 sites and their basins where specimens of *M. galloprovincialis* were sampled

Site	Basin
Palude della Rosa-A	North
Palude Maggiore-B	North
Sant'Erasmo outer-C	North
Sant'Erasmo inner-D	North
Venezia-F	North
Campalto-E	Central
Fusina outer-G	Central
Fusina inner-H	Central
Sacca Sessola-I	Central
Valle Tezze-J	Central
Rivola-K	Central
Valle di Bon-L	Central
Ottagono San Pietro-M	Central
Settemorti-N	South
Pellestrina-O	South
Ca' Roman-P	South
Valle della Dolce-Q	South
Aleghero-R	South
Chiuse di Brenta-S	South

Quantification was performed via microscopic counting [2, 3]. Collected spectra were compared to spectra in different reference libraries by a similarity algorithm. When operating with this software section, the optimal range of match percentage is >65%; thus, only particles with a match percentage ≥65% were considered optimally identified and then quantified.

2.3 Results and Discussion

Preliminary results are presented. Different abundance and polymer distribution were observed (Fig. 2.3).

PA 6 (polyamide 6 or nylon) was the prevalent polymer; however, other polymers such as fluorinated polymers (TFE, PTFE) and olefin fiber were observed. The average length of these μparticles was 55 μm, although few particles of PA 6 had a length above >100 μm and an aspect ratio >3, which means these particles can be considered fibers.

In addition to abundance, which is high, it is also appropriate to consider the weight of SMPs per gram (e.g., the weight of PA 6 in specimens from Campalto was around 10 μg per gram of *M. galloprovincialis*).

Fig. 2.2 A plastic-free clean room where the pre-treatment procedure was performed

Differences were also observed in the distribution and abundance of additives and plasticizers. Some of these compounds have specific uses, i.e., vulcanizing agents that can be proxies of plastic polymers.

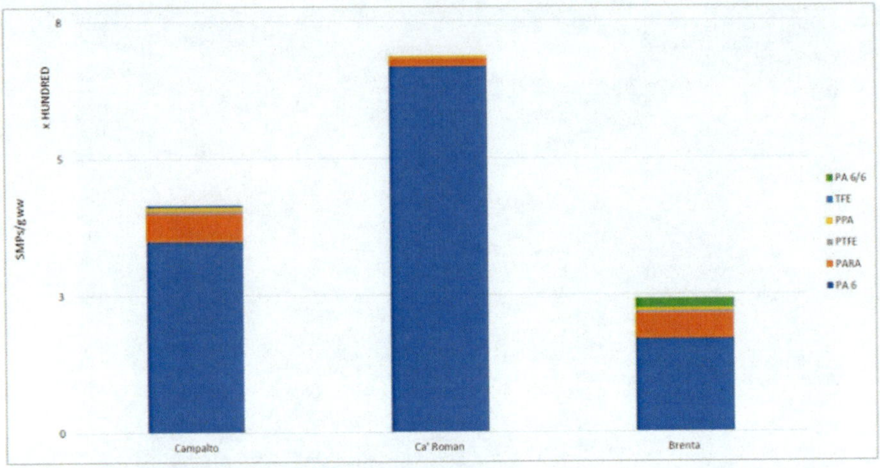

Fig. 2.3 Abundance and distribution of SMPs (5–100 μm) in some of the sites investigated

2.4 Conclusions

These preliminary data will allow an in-depth understanding of the sources and pathways of plastic polymers in the lagoon and its catchment area. The differences observed in abundance, polymer distribution, and weight of SMPs in specimens of M. galloprovincialis can help drawing a map of plastic pollution in the Venice Lagoon. Besides, knowing the distribution of additives, plasticizers, and other micro-litter components will add further details to this map of plastic contamination. Furthermore, these data may be helpful in evaluating the environmental risk assessment and health assessment for humans as potential seafood consumers.

References

1. ECHA (European Chemicals Agency) (2019) Annex XV restriction report proposal for a restriction. Report ver. 1
2. Corami et al (2022) Science of the total environment, p 113477
3. Corami et al (2020) Marine pollution bulletin, p 111606
4. Nelms et al (2019) Scientific reports, p 1075
5. Savoca et al (2021) Global changes biology, pp 2188–2199

Chapter 3
Step-By-Step Evaluation of Matrix Effect and Sample Recovery During Pretreatment of Wastewater Samples for the Quali-Quantitation of Microplastics by Py-GC-MS

Jeanette Lykkemark, Marco Mattonai, Alvise Vianello, Jes Vollertsen, and Francesca Modugno

Sample pretreatment strategies for quali-quantitative analysis of microplastics (MPs) in environmental samples usually entail several separation, digestion, and purification steps, in which plastics particles are isolated from both the inorganic and organic components of the matrix. The aim of such pretreatments is to increase the method sensitivity and reduce the interference of the matrix on the analysis. However, limited information is available on the recovery of plastics during such pretreatments, and on the entity of matrix effects generated by matrix components. A more in-depth knowledge of these aspects could provide valuable insights to further optimize pretreatment strategies.

We describe a systematic investigation of the effects of a sequence of widely used pretreatment steps for quantitative MPs analysis in wastewater samples by analytical pyrolysis-gas chromatography-mass spectrometry (Py-GC-MS). The goal was to evaluate the effect of sample pretreatment in the Py-GC-MS quantification of MPs.

A set of replicate samples from a wastewater treatment plant were subjected to an increasing number of modular different pretreatment steps, and the cumulative effects of these steps on Py-GC-MS recovery and sensibility were evaluated on

J. Lykkemark (✉) · A. Vianello · J. Vollertsen
Department of the Built Environment, Aalborg University, Aalborg, Denmark
e-mail: jly@build.aau.dk

M. Mattonai · F. Modugno
Department of Chemistry and Industrial Chemistry, University of Pisa, Pisa, Italy

J. Lykkemark · M. Mattonai · J. Vollertsen · F. Modugno
North Atlantic Microplastic Centre (NAMC), Bergen, Norway

M. Cocca et al. (eds.), *Proceedings of the 3rd International Conference on Microplastic Pollution in the Mediterranean Sea*, Springer Water,
https://doi.org/10.1007/978-3-031-34455-8_3

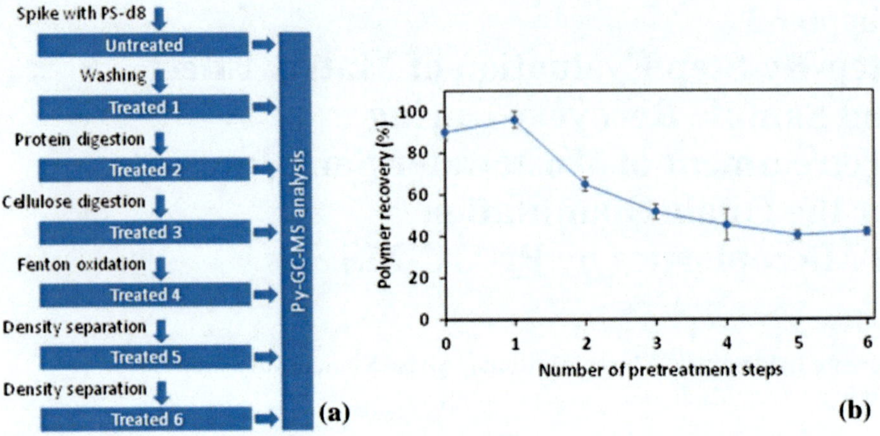

Fig. 3.1 **a** Flowchart of the analytical methodology followed in this study and **b** PS-d8 recoveries evaluated with in-matrix calibration as a function of the number of pretreatment steps

deuterated polystyrene (PS-d8), which was used as internal standard. Samples preparation consisted of six-steps including (1) washing with surfactants, (2) enzymatic digestion of proteins, (3) enzymatic digestion of cellulose, (4) Fenton oxidation, and (5, 6) single and double density separation. After each step, a fraction of the samples was analysed by Py-GC-MS, while the others were brought forward to the next step (Fig. 3.1a). Aliquots of the samples were spiked with different amounts of PS-d8, allowing us to build in-matrix calibration curves at each step of the pretreatment. This allowed us to estimate polymer recoveries and matrix effects throughout the sample treatment process.

The results showed a loss in polymer amount as the number of pretreatment steps increased (Fig. 3.1b). Recoveries as low as 40% were determined for the samples which underwent all pretreatment steps. All in-matrix calibration curves provided good r2 values. Sensitivity and response linearity at each pretreatment step were only marginally affected by the presence of the environmental matrix, suggesting that minimal matrix effect was present.

Multi-step pretreatment strategies can lead to loss in polymer amount, with the risk of analytes falling below the instrumental detection limits. In-matrix calibration could be a promising strategy when dealing with samples with very low polymer amounts, in which an extensive sample pretreatment could cause significant sample loss. Future studies should be aimed at investigating polymer recovery and the effectiveness of in-matrix calibration when dealing with other environmental matrices, such as soil, seawater, and biota.

Acknowledgements The research was carried out in the context of the North Atlantic Microplastic Centre NAMC coordinated by the NORCE Norwegian Research Centre and implemented after a donation of the Agenda Vestlandet (Bergen, Norway).

Chapter 4
Challenges for Restricting Microplastics in the EU: Urgent Need for Recommendations from Scientists on the Restriction Proposal

Esther Kentin

Abstract The European Union (EU) initiated the process of restricting microplastics in 2017, but no restriction has been approved yet. Scientific evidence on microplastics has been rapidly increasing, raising questions about whether the proposed restriction is based on the latest evidence. This chapter discusses two key issues: the definition of microplastics and the exemption for biodegradable microplastics. Does the restriction proposal reflect current scientific consensus? The chapter emphasizes the need for scientists to engage with policymakers and provide bold recommendations. The restriction proposal is currently under review by the European Parliament and the Council, presenting an opportunity to address the remaining questions and concerns.

4.1 Introduction

The process of restricting of microplastics in the EU started in 2017 with a request of the European Commission (EC) to compile a restriction dossier under the European Chemicals regulation (REACH) [1]. The procedure under REACH foresaw an adoption in 2020, but unfortunately no restriction has been approved yet. Our environment is persistently being polluted by microplastics, with emissions seemingly on the rise. As part of the process, the restriction proposal has been deliberated on in several consultation rounds, and the Committee for Risk Assessment (RAC) and the Committee for Socio-economic Analysis (SEAC) have provided opinions and suggestions for revision. Enforcement authorities of the EU Member States have also given input via the Forum for Exchange of Information on Enforcement (the Forum). While the restriction process is slowly moving forward, scientific evidence is building up. A quick—therefore incomplete—search in Web of Science showed a large increase in relevant articles between 2010 and 2020, from four to 1,646. In

E. Kentin (✉)
Institute for the Interdisciplinary Study of Law, Leiden University, Leiden, The Netherlands
e-mail: e.kentin@law.leidenuniv.nl

M. Cocca et al. (eds.), *Proceedings of the 3rd International Conference on Microplastic Pollution in the Mediterranean Sea*, Springer Water,
https://doi.org/10.1007/978-3-031-34455-8_4

2021, there were a further 2,614 articles. The restriction dossier compiled by the European Chemicals Agency (ECHA) was finalized in August 2019, meaning that it does not cover the many more recent studies and results. The documents prepared by RAC and SEAC also do not contain studies from after 2019. This raises the question of whether the proposed restriction can be considered to be based on the latest scientific evidence.

I will discuss two issues in particular: (1) the definition of microplastics; and (2) the exemption for biodegradable microplastics [2]. The definition of microplastics with a lower size limit of 100 nm is defended by SEAC based on practical and technical considerations, such as analytical difficulties, while RAC concludes that the revised definition may lead to circumvention of the restriction via substitution of microplastics with nanoplastics. The second issue relates to the derogation of biodegradable polymers. RAC considers the 'all compartments' testing method is most appropriate, but points out that none of the proposed testing methods would fully address all uncertainties. 'Biodegradable microplastics' may be less degradable in certain environments and therefore not always be eliminated as a source of microplastic pollution.

Can current scientific evidence justify the lower size limit of 100 nm and the exemption for biodegradable microplastics?

4.2 The Restriction Procedure Under REACH

Under REACH, chemicals are regulated with the adage 'no data, no market'. To bring a chemical onto the market, a manufacturer or importer is required to register and supply a chemical safety sheet to the European Chemicals Agency (ECHA). If there is reason to do so, the chemical can be further evaluated, authorised under conditions, or restricted, meaning it can be prohibited totally, or for certain uses. Under Article 68(1) of REACH, restriction is triggered where there is 'unacceptable risk to human health or the environment'.

The restriction procedure under REACH starts with the compilation of a restriction dossier. This can be done by one or more EU Member States, but also by ECHA at the request of the EC, as happened for microplastics. The restriction dossier has a strict format: in short, it should contain identification of the problem, the uses of the chemical, information on its risks, an impact assessment, stakeholder information, assumptions and uncertainties, a justification for EU regulation and, of course, a proposal for restriction. The restriction dossier on microplastics is distinctive in several ways. First, the dossier deals with substances—polymers—that were excluded from registration and thus evaluation under REACH in the first place. When REACH was negotiated, most polymers were considered 'polymers of low concern' and regulatory requirements were reduced [3]. Second, the grouping of so many substances under the definition microplastics is unprecedented. Third, the classification of microplastics as a non-threshold substance is based on persistence and bioaccumulation. Fourth, while compiling the restriction dossier, there has been an enormous increase in scientific research, although many gaps and uncertainties remain. The restriction dossier states that further information should be collected for possible additional measures.

4.2.1 Definition of Microplastics: The Lower Size Limit

The first proposal defined the size of microplastics as follows: (i) all dimensions 1 nm $\leq x \leq 5$ mm, or (ii), for fibres, a length of 3 nm $\leq x \leq 15$ mm and length to diameter ratio of >3. This definition was partly derived from other regulatory initiatives, such as the US Microbead-Free Waters Act and the French prohibition of microplastics in cosmetic products. Furthermore, in scientific literature microplastics were mostly characterized as particles smaller than 5 mm.

After consultations with stakeholders, ECHA revised the lower size limit to 100 nm (or 0.1 μm as described in the proposed restriction) and 300 nm for fibres, therefore leaving out nanoplastics. This revision was motivated by 'obvious analytical constraints and challenges' and 'based on practical and legal certainty'. With growing attention and research on nanoplastics, some researchers suggest that nanoplastics should be excluded from the definition of microplastics, introducing a lower size limit [4, 5]. While this might be useful for research and certain aspects of regulation, such as identification and monitoring, excluding nanoplastics from the restriction proposal raises the question of whether this will undermine the purpose of the restriction.

RAC proposes to fully remove the lower size limit to explicitly include nanoplastics, as nanoplastics may have greater potential than microplastics for adverse effects and bioaccumulation. RAC points to existing use of nanoplastics in cosmetic products and the danger of substitution to circumvent the restriction. A lower limit would make it possible for industry to replace microplastics with nanoplastics. This is not unrealistic, as nanomaterials are already frequently used in cosmetic products. Under the Cosmetic regulation, a manufacturer is required to label nanomaterials in ingredients lists. While nanomaterials, including nanoplastics, may also be used in other products, such as coatings for textiles, their use is relatively unknown as no other labelling requirements apply.

Conversely, SEAC concludes a lower size limit of 1 nm should be applied, however with a temporary lower size limit of 100 nm 'in order to ensure that the proposed restriction is implementable, enforceable and monitorable', because 'the reliable characterisation or identification of microplastics is not self-evident'. These technological barriers should be removed, and ideally as soon as is possible. Furthermore, SEAC considers that use of the generic term 'microplastics' is unfortunate, and that physical and chemical criteria should have been used instead.

Given that these opinions were issued in 2020, are they still reliable? Monikh et al. [6] conclude that, on the basis of progress that has been made, nanoplastics should be included in the restriction proposals [6]. Enforcement may be difficult, like with regulation of nanomaterials in general, but is not impossible. Nonetheless, can it be effective and economical?

4.2.2 The Exemption for Biodegradable Plastics

As the restriction proposal is based on the property of persistence of plastics, biodegradable microplastics are excluded. In the Single Use Plastic directive (SUP) of 2019 there is no exception for biodegradable plastics. Having learned from the debacle of oxo-degradable plastics, the SUP rules also apply to biodegradable plastic products. An evaluation report foreseen in 2027 might include an assessment of biodegradability standards to be included, ensuring 'full decomposition into carbon dioxide (CO_2), biomass and water within a timescale short enough for the plastics not to be harmful to marine life and not to lead to an accumulation of plastics in the environment'. The approach of the restriction proposal is different: it does not refer to harmfulness, but includes a tiered approach using criteria as 'ready biodegradation', 'inherent biodegradation', and 'bio(degradation) relative to a reference material', demonstrating half-life in the environment of 60 or 180 days, under relevant environmental conditions. Criteria refer to OECD and EN ISO standards, and tests have to be conducted by accredited laboratories.

After lengthy consideration of whether 'biodegradable' materials could still contribute to concerns with microplastics, RAC proposed modified criteria to reduce uncertainties related to biodegradability.

In 2020, SAPEA published a report on the biodegradability of plastics, and the Group of Chief Scientific Advisors then published a scientific opinion on this issue. An important conclusion is that biodegradability should be considered as a 'system property' rather than a material property [7]. The example of microplastics in cosmetics highlights the challenges: biodegradable microplastics should degrade rapidly in a range of open environments, as they may be present in sewage sludge, but can also be released directly into freshwater or marine environments. Therefore, a multi-tier test scheme is needed. The Group of Chief Scientific Advisors followed the report and recommends defining biodegradability as a system property [8]. Furthermore, the Group endorses limiting the use of biodegradable plastics for specific applications for which reduction, reuse and recycling are not feasible. This view diverts from the approach in the current restriction proposal which includes a general exemption for biodegradable microplastics, without specifying use or application. Additionally, it takes into account the waste hierarchy approach, embedded in the EU regulation on waste, and the principle of prevention and precautionary principle, the underlying principles of EU environmental policy. Do the current tests in the restriction proposal follow the recommendations made by SAPEA and the Group of Scientific Advisors? Do the standards included in the exemption make up for the system property approach? Is there new scientific evidence that should be considered?

4.3 Discussion

In my opinion scientists could be more proactive, or even assertive, to discuss these concerns and to inform policymakers. Research may could be directed in this direction, in particular since we face a serious and irreversible environmental problem. During the conference many new studies were presented but only few presentations addressed possible policies and regulatory options to reduce the emission of microplastics. The identified questions remain. Does the restriction proposal reflect current scientific consensus? Will the restriction minimise the releases of intentionally added microplastics to the environment as envisaged by ECHA? I addressed only two issues in the restriction proposal: the lower size limit and the exemption for biodegradable microplastics, but there are more concerns to be addressed. The restriction proposal has entered the next phase is now being sent for adoption to the European Parliament and the Council. Both organs can make amendments while the Commission may also make modifications during this process. There is still time to raise and resolve the above questions. I urge scientists to act and make bold recommendations to policymakers in Europe and its Member States.

References

1. ECHA (2019) Annex XV restriction report on intentionally added microplastics Helsinki
2. ECHA (RAC and SEAC) (2020) Opinion on an annex XV dossier proposing restrictions on intentionally-added microplastics Helsinki
3. Kentin E (2018) Restricting microplastics in the European Union: process and criteria under REACH. Phys J Plus 133:425–437
4. Frias JPGL, Nash R (2019) Microplastics: finding a consensus on the definition. Mar Pollut Bull 138:145–147
5. Gigault et al (2018) Current opinion: what is a nanoplastics? Environ Pollut 235:1030–1034
6. Monikh FA et al (2022) Can current regulations account for intentionally produced nanoplastics? Environ Sci Technol 56:3836–3839
7. SAPEA (2020) Science advice for policy by European academies: biodegradability of plastics in the open environment. SAPEA, Berlin
8. Group of Chief Scientific Advisors (SAM) (2020) Biodegradability of plastics in the open environment. EU, Luxembourg

Chapter 5
What Kind of International Legally Binding Instrument on Plastic Pollution Do We Need?

Federica Tommasi and Laura Mancini

5.1 Introduction

For the international challenge on the plastic issue, this will be perhaps, remembered as the hottest spring and the turning point.

Last March the UNEP Assembly [1] finally decided on a historic legally binding intervention on plastic. We finally have the perception that the challenge is not only global, but that solution is approaching. In 2021, the amendments to the Basel Convention on plastics became effective. The very recent COPs of the Stockholm, Basel and Rotterdam Conventions (as we write), has also put plastic at the center again, as an emerging pollutant and with its impact everywhere, not only starting from aquatic environments. After years in which the growing alarm about MPs Pollution has produced an ever-growing literature on environmental sectors and human health, finally all international bodies seem to converge to tackle the problem (EU, G7, OECD, UNEP, WEF, WHO, FAO).

But what will be the premises, the foundations, the indispensable elements but above all the method, which are the elements that cannot be missing, to be able to say that the effort made becomes a success?

Surely we need a simple, clear, gradual, and shared agreement that on a global level starts from the existing regulatory framework, confirming what already binds the international community, expanding its audience. More than big unattainable resolutions, this phase of initiating the work of the delegate group requires clear input and method.

Such a far-reaching agreement must be shared, understandable also for the societies and stakeholders who must accept it, because it has great implications, not only

F. Tommasi (✉) · L. Mancini
Ecosystems and Health, Department Environment and Health, ISS – Italian National Institute of Health, Viale Regina Elena, 299, 00161 Roma, RM, Italy
e-mail: federica.tommasi@iss.it

M. Cocca et al. (eds.), *Proceedings of the 3rd International Conference on Microplastic Pollution in the Mediterranean Sea*, Springer Water,
https://doi.org/10.1007/978-3-031-34455-8_5

economic, but also sociological and cultural. Will we be able to take up this challenge, inevitably linked to the others we have faced recently, such as the pandemic and climate challenges?

We have only one possible destiny: success with the reversal of the trend in consumption and lifestyles linked to the abuse of plastic materials. Defeat is actually our present day that we need to change.

5.2 The Way Forward a New Treaty on Global Plastic Pollution

A 2022 full of steps forward on Marine Litter and Plastic Pollution Initiatives, Laws and new Agreements. United Nations General Assembly recognizes as a Human Right a clean, healthy, and sustainable Environment for all mankind [2].

The UN General Assembly (UNGA) passed on 28 July 2022 a Resolution (A/76/L.75) recognizing the right to a clean, healthy, and sustainable environment as a human right. The UNGA calls upon Parties and other stakeholders to "scale up efforts" to ensure a clean, healthy, and sustainable environment for all.

Quoting that the right to a clean, healthy, and sustainable environment is "*related to other rights and existing international law,*" and affirming that its promotion "requires the full implementation" of the Multilateral Environmental Agreements (MEAs) "under the principles of international environmental law." These general principles paved the road ahead also tackling the plastic pollution globally with a new dedicated international Treaty (Fig. 5.1).

This result has been achieved during the session known as UNEA 5.2 convened in Nairobi, Kenya from 28 February-2 March 2022, a year after the virtual fifth Assembly. The UNEA is composed of the 193 UN Member States, during its closing session, governments adopted a resolution titled, 'End Plastic Pollution: Towards an internationally legally binding instrument.' By Resolution UNEP/EA.5/Res.14 the Assembly agrees to set up an Intergovernmental Negotiating Committee (INC) to draft a Legally Binding Agreement by 2024. The agreement is expected to address the full lifecycle of plastics, including production, design, and disposal, as well as the design of reusable and recyclable products and materials. This represents a great point of arrival, and a challenge for the future, after years marked by a growing global commitment on the Plastic pollution front. Both at the level of world organizations and of global commitments, the issue of Plastic Pollution was now on the agenda of international tables (G7 Germany on 2022, G20 with the Osaka Blue Ocean Vision, The G7 with the Ocean Plastics Charter) [3, 4].

Also in the past years the Plastic Pollution and the Marine Litter Pollution are addressed also in the work of: Basel Convention on the Control of Transboundary Movements of Hazardous Wastes and Their Disposal (Basel Convention), Strategic Approach to International Chemicals Management (SAICM), International Maritime Organization (IMO), World Health Organization (WHO), World Trade Organization

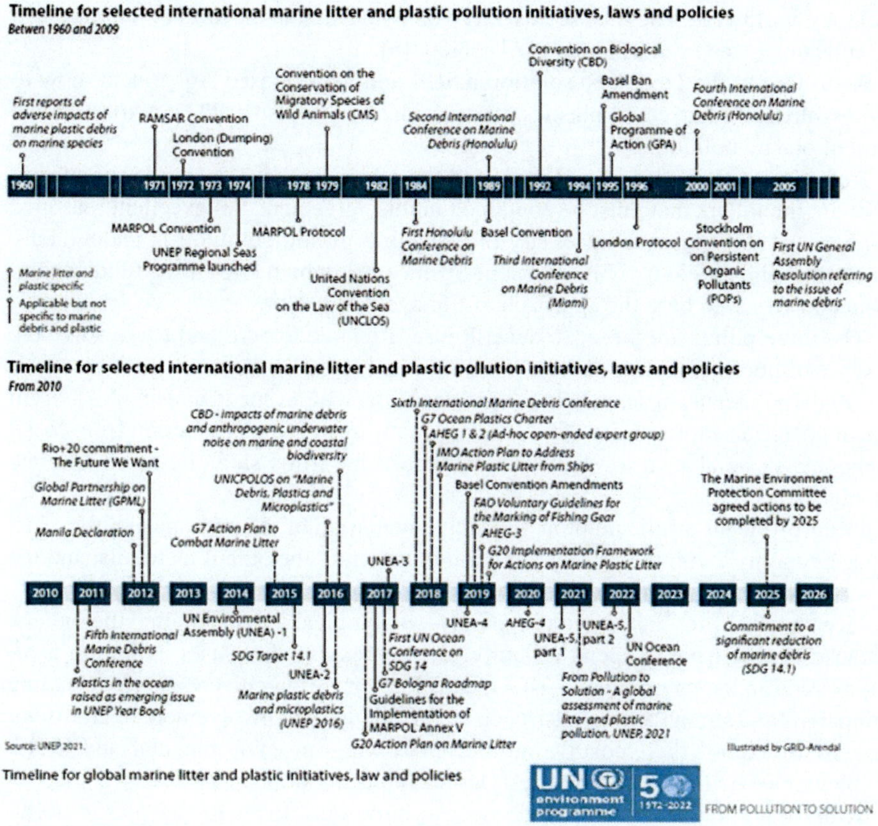

Fig. 5.1 Timeline for global marine litter and plastic initiatives, law and policies. *Source* UNEP [3]

(WTO), Various Regional Seas Programmes and Conventions around the world. Therefore a great transversal commitment of global institutions for what has been defined as one of the contaminations characterizing the new era of the Anthropocene.

But if at an institutional level the course now seems to have been set, how ready is civil society to accept a new treaty? We must consider that it will also have an impact on its lifestyles and consumption, regardless of the economic issues that this type of transition implies with a lower and sustainable consumption of plastic.

For sure quite enough if we look at the results of the survey of Dilkes-Hoffman [5] carried out among Australian citizens who were asked to indicate which, amid the various proposals, was the most serious environmental problem. Also surprisingly, it has been answered at the top of the list: "Pollution in the ocean" even before more urgent and dangerous issues as, water pollution, air pollution, climate change and global warming. This, like other published surveys, demonstrates the great interest, concern and perception of the impact of plastic pollution in developed societies worldwide. The road seems paved on both sides: at the level of international organizations, and on the side of civil society [6].

Returning to the UNEA Resolution n.5/14 and its proposed Agreement, how to address the full lifecycle of plastics, from production to disposal? How to reach the "end of plastic pollution"?

Before to explain that crucial point and the roadmap ahead, let try to summarize what are the pillars that must be contained in this agreement. An excellent reasoned synthesis of the approach necessary in managing plastic pollution is undoubtedly reported in the article by Nils Simon in Science [7], which cites three fundamental pillars on which to base the approach of life cycle of plastics.

The three pillars are these: Reduce, Reuse, Remove to curb and try to solve the plastic pollution problem.

"Reduce" means, in a nutshell, the greater effort we can spend in using less virgin plastic material starting from the S.U.P. products, to curb the production from fossil fuels, and to be able to lower GHG emissions, due to its side effects on Climate Change [8].

"Reuse" means a full and proactive implementation of the principles of the "Circular Economy" with the greater attention in greening the output materials, and the related production processes [9].

"Remove" represents the more purely sociological issue, among the others, because we too have to spend all our efforts to remove the plastic pollution actually present in the environment. This means implementing citizen science, cleaning campaigns and strengthening all aspects of knowledge and involvement of consumer citizens to exploit their positive attitude in solving an environmental and health problem, closely linked to their lifestyles and consumption.

If these are the pillars, on which we can build our efforts to end plastic pollution, how do we achieve these results in a general and global agreement? On which obviously, in a general and shared way, all member states must agree.

In understanding how this process will develop, first of all, we have to resume the rules in doing that effort: an Intergovernmental Negotiating Committee (INC) will write the new Agreement; and an Ad-hoc Open Ended Working Group (AH-OEWG) has decided the "Rules of Procedure" and "Meeting Schedule" of INC, as prescribed in the Res.5/14, establishing the framework for negotiations and the details matters, within the ambitious deadline of the end of 2024.

Such a stringent deadline, previously unthinkable, for other agreements already signed, testifies to the fact that for the framework we want to give to this new Agreement, we want to draw inspiration from recent and past experiences. "Minamata Convention" [10] has been see as a good general Roadmap to develop a new International Legally Binding Instrument (ILBI) on Plastic Pollution.

Indeed precisely this recent Convention deals with the issue of Mercury pollution, managing, with progressive prohibitions, both the production processes, in which Mercury is used, and the products that contain it. In this regard, this Convention also provides for the negotiation methods with which to manage the economic instruments and the progressive phasing-out of products containing mercury.

This convention represents the best skeleton on which to mend the new treaty for plastic pollution which precisely requires the combination of the progressive ban of the most polluting materials and products, and the related economic instruments necessary to achieve these objectives.

From one side for its procedures to apply in negotiations, and on the other for its model in approaching the new plastic pollution treaty with controls and scheduled phasing-out on various plastic products (*Reduce: Goal 1*).

In addition, in the first meetings of the Ad-hoc Open Ended Working Group it emerged that also Paris Agreement too has been suggested, to replicate somehow, its NDCs Nationally Determined Contributions (biggest polluters put in place the biggest actions). In the debate, also outside the OEWG held in June, raised the proposal of a sort of a "Cap and trade" Principle both for virgin and/or waste streams reinforced by a taxation on virgin plastic production (*Reuse: Goal 2*) e.g. at the OECD Workshop: "FFGP—Economic, Regulatory or Technical Barriers to Sustainable Design from a Chemicals Perspective—How Can Policy Makers Help?" hold in May of 2022.

In other word therefore, the best solution is to give a proper global response that has to address both virgin plastic production and waste production in a double fast track, addressing global plastic pollution with a comprehensive life cycle approach [11].

5.3 Conclusions

We need to take action, anyway, at any cost, with the utmost willingness to negotiate globally, for such an ambitious goal, in such a short time, paving a way with an "open and incremental and implementable Legally Binding Instrument".

Because if for the "common man" the plastic problem seems to be limited only to the issue of packaging containment, the real future, managing the plastic pollution, will be represented by everything else made by plastic materials—also plastic items with legacy chemicals content—you are able to imagine, to be managed till the remote future.

Resolution 5/14 demonstrates that inaction is not an option and that the wrong actions are those of today: the sustainable future, even for an intelligent and versatile material such as plastic, must be thought of today, with a global negotiation, because this problem is global.

Such a close deadline (2024 is a written challenge for next year!) also demonstrates that the problem is urgent and globally perceived as such.

References

1. United Nations Environment Assembly of the United Nations Environment Programme. Resolution adopted by the United Nations Environment Assembly on 2 March 2022 5/14. End plastic pollution: towards an international legally binding instrument. https://wedocs.unep.org/bitstream/handle/20.500.11822/39812/OEWG_PP_1_INF_1_UNEA%20resolution.pdf
2. United Nations General Assembly: Seventy-sixth session Agenda Item 74 (b). The human right to a clean, healthy and sustainable environment: resolution/adopted by the General Assembly. https://digitallibrary.un.org/record/3983329?ln=en
3. United Nations Environment Programme (2021) From pollution to solution. A global assessment of marine litter and plastic pollution Nairobi. https://www.unep.org/resources/pollution-solution-global-assessment-marine-litter-and-plastic-pollution
4. Government of Canada: Ocean Plastics Charter (2018). https://www.canada.ca/en/environment-climate-change/corporate/international-affairs/partnerships-organizations/ocean-plastics-charter.html
5. Dilkes-Hoffman LS (2019) Public attitudes towards plastics. Res Cons Rec 147:227–235
6. Stokstad E (2022) United Nations to tackle global plastics pollution. Science 375(6583):801–802
7. Simon N (2022) A binding global agreement to address the life cycle of plastics. Science 373(6550):43–47
8. Bergmann M (2022) A global plastic treaty must cup production. Science 376(6592):469–470
9. Grosso M (2022) It's all about plastics. Waste Man Reas 40(6):607–608
10. Minamata Convention. https://www.mercuryconvention.org/en
11. Earth Negotiations Bulletin. Summary of the Ad hoc Open-ended Working Group to Prepare for the Intergovernmental Negotiating Committee to Develop an International Legally Binding Instrument on Plastic Pollution, including in the Marine Environment: 30 May—1 June 2022. https://enb.iisd.org/sites/default/files/2022-06/enb3601e.pdf

Chapter 6
Detection and Characterization of Micro- and Nanoplastic Pollutants: A Case Study of the Adriatic Sea

Maria Râpă, Annamaria Vujanović, Teresa Cecchi, and Davide Poletto

Abstract This study investigates the detection of microplastic in the Adriatic Sea, considering various plastic waste-generating activities from neighboring countries as a source of pollution. Seven water samples were collected from three locations around the Venice Lagoon in Italy and four around the island of Krk in Croatia. Microplastics, categorised as fibres, films, and fragments, were quantified, and their features (dimensions, colour) were analysed via optical microscopy. The mean size, undersize (D10, D50, and D90), and concentration of nanoparticles were also estimated using Nanoparticle Tracking Analysis. The Flame Atomic Absorption Spectrometry (FAAS) also scutinized the presence of certain heavy metals in the water bodies after the digestion step. This scientific investigation reveals the highest concentrations of microplastics and nanoparticles in the Canal Grande at the city centre near the Rialto Bridge, one of the most iconic and massively visited locations in Venice. Finally, the levels of heavy metals in the seawater were below the detection limit of FAAS. This study paves the way for assessing the impact of these pollutants on aquatic ecosystems and on human health.

M. Râpă
Faculty of Materials Sciences and Engineering, University Politehnica of Bucharest, Bucharest, Romania

A. Vujanović
Faculty of Chemistry and Chemical Engineering, University of Maribor, Maribor, Slovenia

T. Cecchi
Chemistry Department, Istituto Tecnico Tecnologico (ITT), Fermo, Italy

D. Poletto (✉)
Venice Lagoon Plastic Free, Venice, Italy
e-mail: d.poletto@plasticfreevenice.org

M. Cocca et al. (eds.), *Proceedings of the 3rd International Conference on Microplastic Pollution in the Mediterranean Sea*, Springer Water,
https://doi.org/10.1007/978-3-031-34455-8_6

6.1 Introduction

In the EU, 150,000–500,000 tons of plastic waste enter the oceans every year with 75,000 and 300,000 tons of microplastics released into the environment. Microplastics (MPs) are defined as plastic particles resulting from the disintegration of plastics, with dimensions smaller than 5 mm, which are not detectable to the naked eye [1]. Relating to the direct release into the environment or degradation of plastic waste, MPs can be categorized as primary and secondary microplastics. Laboratory research suggests that the plastic pollution of the oceans, seas and lakes has induced harm to aquatic life. MPs have been discovered in biota, such as marine algae, crabs and whales, or comestible salts [2]. Also, more than 701 marine species have ingested these microplastics [3]. Fish and small organisms can accidentally eat the plastic particles, leading to reduced fertility, decreased energy and altering juvenile development of invertebrates. Furthermore, the plastic pollution of marine environments has raised significant concerns relating to human health due to the chemical components of plastic being linked to serious diseases, including cancer.

Water pollution is also a result of naturally occurring contaminants caused by agriculture, and household and industrial waste that can contaminate MPs through sorption processes. Among these, heavy metals are the most dangerous contaminants due to their toxicity and accumulation in organic organisms. They are able to bioaccumulate in the food chain, thus influencing ecosystems and, ultimately, human health.

This negative impact on the environment is widely recognized as being unacceptable at a biological, ecological and socio-economic level [4]. Consequently, aquatic ecosystems have been threatened by an exponential increase in plastic debris; its impact is not entirely understood as an overall systemic crisis of our oceans, driven by habitat degradation, loss of biodiversity, climatic transformation and alteration and disruption of major sea streams [5].

This study investigates the occurrence and characterization of microplastic contamination in the Adriatic Sea, where different plastic waste generating activities from neighbouring countries are considered as a pollution source. According to the applied protocol, water samples have been collected in two rounds from three locations around Venice Lagoon in Italy and four locations around the Island of Krk in Croatia. Microplastics, perceptually categorized as fibres, films, and fragments, were quantified for each water sample; their morphological features (dimensions, colour) were studied using optical microscopy. The mean size, undersize (D10, D50 and D90) and concentration of nanoparticles were also estimated using a Nanoparticle Tracking Analysis technique. Furthermore, heavy metals have been quantified in the bodies of water after the digestion step by flame atomization using an atomic absorption spectrometry analytical tool.

Table 6.1 Water samples and sampling areas

Sample code	Location	GPS coordinates	
		Latitude (°)	Longitude (°)
Venice Port Inlet	Venice, Italy	45.431508	12.406952
Rialto Bridge	Venice, Italy	45.438350	12.336311
Saint Mark's Basin	Venice, near St Mark's Square, Italy	45.431962	12.340953
Stara Baška	Island of Krk-Stara Baška-beach Zala, Croatia	44.948901	14.699455
Krk	Krk, Croatia	45.012987	14.561584
Melska	Island of Krk-Melska, Croatia	45.100624	14.661386
Omišalj	Omišalj, Croatia	45.244462	14.554328

6.2 Experimental

6.2.1 *Water Sampling and Storage*

Three bulk water samples were collected from Venice, Italy, and four others from Croatia (Table 6.1) on 7th December 2021 (for nanoparticles and heavy metals analyses) and 15th June 2022 (for microplastics determination) (30 cm under surface) using stainless steel jerry cans.

According to our sampling protocol, the containers were rinsed at least 4 times with seawater at each sampling location and allowed to settle for 5 min between each rinse. The water samples were kept at a cold temperature (8 °C ± 2 °C) during transportation to the laboratory. The samples were stored in refrigerators at the same temperature and processed two days after being received.

6.2.2 *Methods*

6.2.2.1 Laboratory Preparation

During water processing in the laboratory, some precautions were taken, including wearing a cotton laboratory coat and nitrile gloves. Prior to the commencement of work, the basic laboratory glassware, Petri dishes, mesh sieves, filtration system, as well as work surfaces were cleaned with 70% ethanol, then rinsed 3 times with filtered ultrapure water and covered with aluminium foil.

In order to avoid contamination from particles or experimental reagents, airborne contamination was monitored according to the procedure of Wang et al. [6] and three blank samples (negative controls) were performed simultaneously in each analytical experiment. Upon completion of the work, the filter papers were examined for the presence of microplastics (MPs) using an optical microscope (BX 51 M, Olympus, USA).

All working solutions (ultrapure water, ethanol, oxidizing agent) were pre-filtered before use to exclude microplastic contamination from the chemicals by glass microfiber filter papers, 47 mm, 0.45 μm.

6.2.2.2 Laboratory Processing for Microplastic Determination

The water samples were processed in triplicate in compliance with reported protocol [7], with a few modifications.

In this way, 1L of water was filtered with a 32 μm stainless-steel sieve (Test sieve, ISO 3310-1, Germany). The retained solid particles on the stainless-steel sieve were rinsed with filtered ultrapure water and carefully transferred into glass beakers. Subsequently, the resulting mixture was submitted to a digestion step to destroy the organic matter at room temperature using a 30 ml mixture of 30% (w/v) solution of hydrogen peroxide (H_2O_2) (Chimreactiv SRL, Bucharest, Romania), an environmentally benign oxidant, until no foam was observed. The non-toxicity of the reduction product (water) is noteworthy from an environmental standpoint. The involvement of the hydroxyl radical (HO·) contributed to the organic matter oxidation due to its strong oxidation capacity (standard potential 2.80 V versus standard hydrogen electrode [8, 9]). The use of hydrogen peroxide avoided the hydrolysis and degradation of microplastics made of polyesters and polyamides that would occur using NaOH and HCl [10]. Subsequently, the resulting mixture was subjected to vacuum filtration (UniVACUUPUMP 1, Lab Logistics Group GmbH, Meckenheim, Germany) equipped with a suction filtration unit (funnel 250 mL, filter holder NS40/35, filter flask 1000 mL/2000 mL, filter plate 47 mm, porosity 2 and clamp), and glass microfiber filters, binder free with pore sizes of 1.6 μm GF/A (Whatman, China). Filters were stored in labelled Petri dishes, dried at room temperature, and kept in desiccators until the optical microscope investigation. The concentration of microplastics (n/L), their category according to size class (0–100 μm, 101–500 μm; 501–1000 μm, 1001–2000 μm, 2001–3000 μm and 3001–4000 μm), and their colour and shape (sphere, film fragment, and fibre) were recorded. Filtered distilled water was processed as blank, in triplicate, in the same way as the water samples.

6.2.2.3 Nanoparticle Measuring

Nanoparticle tracking analysis (NTA) for water samples was carried out using a Nanosight NS300, NTA 3.4 Build 3.4.4 version (Malvern Instruments, UK), with the permission of INOE2000. The mean size, undersize (D10, D50 and D90) and concentration of particles/mL from water are expressed as the mean of five repetitions ± standard errors. Prior to measurement, 2 mL of original water samples were filtered by 200 nm syringe filter paper (ChromafilR Xtra PES-20/25, Germany). As reference, a standard latex NTA4089 (Malvern United Kingdom) with a size dimension of 200 nm (initial concentration of $3.41 \times 10^9 \pm 3.63 \times 10^8$ particles/mL) was employed as received and diluted with a sterile solution of NaCl (Zentiva SA,

Bucharest, Romania) to a final concentration of $3.42 \times 10^8 \pm 2.78 \times 10^7$ particles/mL.

6.2.2.4 Heavy Metals

Quantitative analysis for Cu, Mn, Zn and Fe ions were performed via Flame Atomic Absorption Spectroscopy (FAAS) (ContrAA 800, Analytik Jena AG, Germany) on digested samples. A blank (control) was processed as water samples. The heavy metals were quantified in the following conditions: a mixture of acetylene-air, read time of 3 s, the washing time between samples of 8 s, and the delay time of 8 s.

Calibration lines were prepared using at least 5 working standards in concentration ranges of 0–2.5 mg/L for Cu, and 0–3 mg/L for Mn, Zn, and Fe (Scharlab SL, Sentmenat, Spain). The LODs of the Cu, Mn, Zn and Fe ions were: 0.0703, 0.1885, 0.1009 and 0.3416 mg/L, respectively.

6.3 Results

6.3.1 Microplastics

The material on each glass microfibre filter was analysed in terms of MPs concentration, category, colour, and shape, detailed in Fig. 6.1a–d.

6.3.2 Nanoparticles

Table 6.2 illustrated the profile of concentration of nanoparticles, mean size and D10, D50 and D90 measured by the NTA technique. D10, D50 and D90 are the points in the size determinations where 10%, 50% (median) and 90% of the sample are found [11], and the agglomeration after filtration can not ruled out. A particle size distribution profile was obtained for each visible particle, which is separately tracked.

6.3.3 Heavy Metals

The concentrations of Cu, Mn, Zn and Fe ions, in ppm, for digested water samples, indicated that the investigated trace metals are below the detection limit of FAAS.

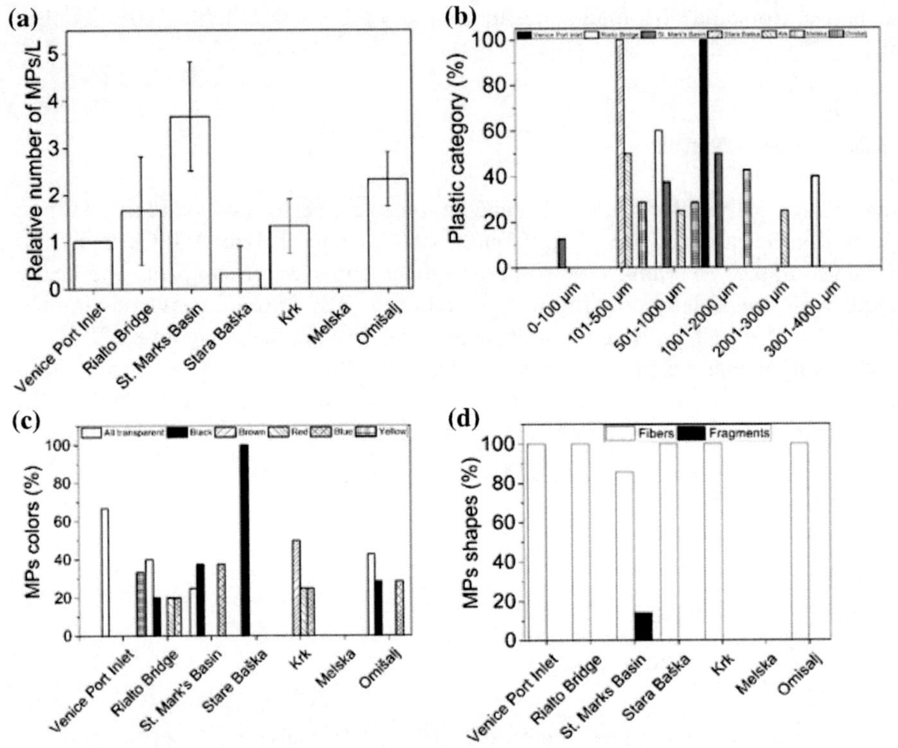

Fig. 6.1 Identification of MPs collected after processing of retained solid particles. **a** Concentration, **b** Category, **c** Colour, and **d** Morphology

Table 6.2 NTA parameters for analysed water samples

Water sample	Mean (nm)	D10 (nm)	D50 (nm)	D90 (nm)	Concentration (particles/mL)
Venice Port Inlet	221.7 ± 59.2	87.9 ± 23.1	132.3 ± 33.6	424.9 ± 117.6	$3.64 \times 10^7 \pm 1.06 \times 10^7$
Rialto Bridge	30.5 ± 18.8	28.7 ± 17.8	30.8 ± 19.0	32.9 ± 20.2	$1.83 \times 10^6 \pm 1.64 \times 10^6$
Saint Mark's Basin	126.4 ± 9.8	87.6 ± 2.6	128.2 ± 12.3	153.1 ± 13.6	$1.82 \times 10^7 \pm 2.60 \times 10^6$
Stara Baška	163.5 ± 35.7	96.9 ± 16.7	146.2 ± 34.7	295.3 ± 65.1	$4.39 \times 10^6 \pm 2.49 \times 10^6$
Krk	85.8 ± 21.6	74.5 ± 18.9	88.6 ± 23.0	98.9 ± 25.5	$9.79 \times 10^6 \pm 2.50 \times 10^7$
Melska	136.0 ± 4.4	85.7 ± 3.5	125.6 ± 14.1	241.0 ± 35.0	$1.87 \times 10^7 \pm 3.61 \times 10^6$
Omišalj	272.7 ± 26.0	136.4 ± 10.8	263.7 ± 30.3	431.0 ± 51.2	$1.25 \times 10^7 \pm 2.17 \times 10^6$

6.4 Discussion

Optical microscopy with the acquisition and image analysis software, and magnification ranging from 50x up to 1000x is used as the first step in the identification of microplastics. It can allow us to distinguish microplastics from other compounds such as fats, mineral particles and cellulose fibres [12]. In this study, three criteria were established for the recognition of MPs as follows: MPs must not have visible cellular or organic structures; fibres should be equally thick throughout their entire length; and particles should present a homogeneous colour throughout [13].

Figure 6.1a shows the relative number of MPs per L in the following trend: Saint Mark's Basin (3.6 ± 1.1) > Omišalj (2.3 ± 0.5) > Rialto Bridge (1.6 ± 1.1) > Venice Port Inlet (1 ± 0) > Krk (1.3 ± 0.5) > Stara Baška (0.3 ± 0.5) > Melska (0). Our results are consistent with data reported in the literature which show the average microplastic abundances in seawater as 9.5 ± 3.7 particles/L [14]. No microplastics were detected on the control samples.

Relative size fraction plotted in Fig. 6.1b shows that the lowest microplastics with diameters up to 100 μm are found in Saint Mark's Basin (12.5%), while the most microplastics in the 3001–4000 μm size class were detected for Rialto Bridge site (40%).

The dominant colours of MPs were transparent, black, and blue (Fig. 6.1c), while the morphology evaluation of MPs showed that the fibres are more predominant than other shapes (Fig. 6.1d).

In terms of nanoparticles, interference from co-existing colloids is the weakness of the NTA technique [15]: actually, they may be nanoplastics but could also be other organic and inorganic nanomaterials. The mean size of the nanoparticles follows the trend: Omišalj > Venice Port Inlet > Stara Baška > Melska > St. Mark's Basin > Krk > Rialto Bridge (Table 6.2).

However, the concentration of nanoparticles found in one millilitre of water sample was situated in the range of $1.83 \times 10^6 \pm 1.64 \times 10^6$ particles/mL to $3.64 \times 10^7 \pm 1.06 \times 10^7$ particles/mL and revealed the following trend: Venice Port Inlet > Melska > St. Mark's Basin > Omišalj > Krk > Stara Baška > Rialto Bridge (Table 6.2).

In aquatic systems, the heavy metals are toxic to organisms above distinct threshold concentrations, but many of them (e.g. Cu and Zn) are fundamental for metabolism at lower concentrations [16]. The quantitative analysis for detection of trace elements in water samples revealed that the resultant spectral peaks are situated under the detection limit of FAAS.

6.5 Conclusions

The results of this scouting scientific investigation show the presence of microplastic particles in all sampling locations except the Melska site; the low volume of sampled water calls for further investigation. The highest concentrations were observed in the Canal Grande in the city centre near Saint Mark's Basin, which is one of the most iconic and widely visited locations in Venice.

Acknowledgements M.R., and D.P. would like to acknowledge the financial support from the H2020 project IN NO PLASTIC (Innovative approaches towards prevention, removal, and reuse of marine plastic litter) https://www.innoplastic.eu/ under the call CE-FNR-09-2020)).

The authors kindly thank Dr. Dean Kosić from PONIKVE eko otok Krk d.o.o, Croatia for providing the water samples and to Dr. Elfrida Cârstea (National Institute for Research and Development in Optoelectronics, INOE 2000) for access to the NANOSIGHT NS 300 tool.

References

1. Wang DJ, Su LC, Ruan HD, Chen JJ, Lu JZ, Lee CH, Jiang SY (2021) Quantitative and qualitative determination of microplastics in oyster, seawater and sediment from the coastal areas in Zhuhai, China. Mar Pollut Bull 164
2. Yaranal NA, Subbiah S, Mohanty K (2021) Identification, extraction of microplastics from edible salts and its removal from contaminated seawater. Environ Technol Innov 21
3. Yap VHS, Chase Z, Wright JT, Hurd CL, Lavers JL, Lenz M (2020) A comparison with natural particles reveals a small specific effect of PVC microplastics on mussel performance. Mar Pollut Bull 160
4. A european strategy for plastics in a circular economy. https://ec.europa.eu/environment/circular-economy/pdf/plastics-strategy-brochure.pdf. Accessed 27 July 2022
5. Irfan S, Alatawi A (2019) Aquatic ecosystem and biodiversity: a review. Open J Ecol 9:1–13
6. Wang WF, Ndungu AW, Li Z, Wang J (2017) Microplastics pollution in inland freshwaters of China: a case study in urban surface waters of Wuhan, China. Sci Total Environ 575:1369–1374
7. La Kanhai DK, Officer R, Lyashevska O, Thompson RC, O'Connor I (2017) Microplastic abundance, distribution and composition along a latitudinal gradient in the Atlantic Ocean. Mar Pollut Bull 115(1–2):307–314
8. Kim K, Lee K, So S, Cho S, Lee M, You K, Moon J, Song T (2018) Fenton-like reaction between copper ions and hydrogen peroxide for high removal rate of tungsten in chemical mechanical planarization ECS. J Solid State Sci Technol 7:91–95
9. Duojia W, Lingcheng Su, Huada DR, Jiajun C, Jianzun L, Chiu-Hong L, Jiang SY (2021) Quantitative and qualitative determination of microplastics in oyster, seawater and sediment from the coastal areas in Zhuhai, China. Mar Pollut Bull 164:112000
10. Poblete-Chavez R, Cortes-Pizarro E, Rojas-Araya R (2016) Treatment of seawater for rotifer culture uses applying adsorption and advanced oxidation processes. Lat Am J Aquat Res 44:779–791
11. Comfort N, Cai K, Bloomquist TR, Strait MD, Ferrante AW Jr, Baccarelli AA (2021) Nanoparticle tracking analysis for the quantification and size determination of extracellular vesicles. J Vis Exp 28:169. https://doi.org/10.3791/624472021
12. Wang ZM, Wagner J, Ghosal S, Bedi G, Wall S (2017) SEM/EDS and optical microscopy analyses of microplastics in ocean trawl and fish guts. Sci Total Environ 603:616–626

13. Cutroneo L, Reboa A, Besio G, Borgogno F, Canesi L, Canuto S, Dara M, Enrile F, Forioso I, Greco G et al (2020) Microplastics in seawater: sampling strategies, laboratory methodologies, and identification techniques applied to port environment. Environ Sci Pollut Res Int 27:8938–8952
14. Zhou Z, Wan L, Cai WQ, Tang J, Wu ZJ, Zhang KD (2022) Species-specific microplastic enrichment characteristics of scleractinian corals from reef environment: insights from an in-situ study at the Xisha Islands. Sci Total Environ 815
15. Zhang W, Wang Q, Chen H (2022) Challenges in characterization of nanoplastics in the environment. Front Environ Sci Eng 16
16. Ansari TM, Marr IL, Tariq N (2004) Heavy metals in marine pollution perspective–a mini review. J Appl Sci 4:1–20

Chapter 7
Microplastics Determination with μ-Raman: Potential and Practical Aspects

Margherita Barchiesi, Camilla Di Marcantonio, Agostina Chiavola, Maria Rosaria Boni, Alessandro Frugis, Valentina Gioia, and Marco Lazzazzara

7.1 Introduction

Drinking water is the result of a chain of treatments aimed at purifying the water source from normed pollutants. However, thanks to the increasing analytical power of the available techniques, more pollutants of concern are being discovered both in the water source and in drinking water. In order to keep the drinking water as safe as possible, such pollutants need to be identified in terms of associated risk and fate in the drinking water treatment plants (DWTPs).

Microplastics (MPs) belong to those pollutants referred to as "of emerging concern". MPs are broadly defined by the scientific community as plastic particles with main dimensions bigger than 100 nm (or 1 μm) and lower than 5 mm [1].

This broad definition coupled with the MPs physical and chemical characteristics results in a challenging task for regulators worldwide for the establishment of acceptable limits and also of standardized analytical protocols.

μ-Raman is among the most used techniques for the analysis of particles as low as 1 μm; however, many are the aspects to be taken into consideration for a reliable and meaningful analysis using this technique. Furthermore, when developing an analytical method, it has to be considered the different aims and goals of a water company, which is responsible for monitoring and managing the water cycle, and of the researchers interested in pure investigation. In the present work, the practical

M. Barchiesi (✉) · C. Di Marcantonio · A. Chiavola · M. R. Boni
Department of Civil, Building and Environmental Engineering (DICEA), Sapienza University of Rome, Via Eudossiana 18, 00184 Rome, Italy
e-mail: margherita.barchiesi@uniroma1.it

A. Frugis · V. Gioia · M. Lazzazzara
ACEA ELABORI SpA, Via Vitorchiano 165, Rome, Italy

M. Cocca et al. (eds.), *Proceedings of the 3rd International Conference on Microplastic Pollution in the Mediterranean Sea*, Springer Water,
https://doi.org/10.1007/978-3-031-34455-8_7

struggles of MPs analysis by μ-Raman are considered, underlining the main issues and proposing solutions to overcome such troubles. It focuses on the activities run in the laboratory, specifically MPs pre-treatments, separation, identification, quantification and characterization (IQC). Firstly, the main steps and issues to carry in the μ-Raman analysis are presented and described; then, the results obtained and the possible solutions are reported.

7.2 Experimental

7.2.1 Materials and Methods

Drinking water for MPs analysis was sourced from private house taps, municipal water towers and DWTPs. Samples were collected in glass containers (volumetric flasks) with volume of 250–1000 mL.

Hydrochloric acid (HCl), acetic acid (AA) and ethanol were of analytical grade and supplied by Carlo Erba, EDTA powder was purchased from VWR.

Silicon filters with porosity of 1–5 μm were used for vacuum filtration, in all glass set-ups. All filtration procedures were run under a laminar hood.

μ-Raman DXR3 by Thermo Fisher coupled with Ominc Spectra and Omnic Atlμs softwares were used for the characterization and identification of MPs.

7.2.2 Pre-treatments

Pre-treatments phase is crucial to a proper IQC. Indeed, without pre-treatments the time needed for the analysis increases, as the particles aggregation and the filter crowding can occur.

Dealing with drinking water, pre-treatments are mostly needed to reduce the presence of Calcium and Magnesium carbonates. At pH of 3, most carbonates salts are supposed to dissolve, being the carbonates mostly present in the form of carbonic acid. Therefore, HCl was used to reach a pH of 3 or lower, whereas trials were also run with acetic acid (AA) reaching a pH of 4.

Another option to dissolve salts is the use of a chelating agent. In the present case, EDTA was selected: it was first dissolved in MilliQ water and then used in an amount equimolar to the carbonates supposedly present in the water analysed.

The experimental conditions tested are shown in Table 7.1

Table 7.1 Experimental conditions for pre-treatment evaluation

Chemical	Value/Dosage	Time
HCl	pH 1–4	Few minutes—12 h
AA	pH 4	Few minutes—12 h
EDTA	Equimolar to carbonates	Few minutes—12 h

7.2.3 Filtration

Filtration was the chosen method for MPs separation. The vacuum filtration system was all glass made, except for the rubber gasket. The silicon filter was squared with a length of 10 mm, whereas the gasket internal diameter was of 9 mm. This introduced a first challenge: optimal positioning of the gasket to avoid water dragging and loss of particles. Issues related to filtration are:

- time needed;
- formation of bubbles due to the high concentration of carbonic acid, its equilibrium with dissolved CO_2 and the local difference in pressure at the pores. The formation of bubbles causes a non-homogeneous distribution of the particles on the filters that strongly hinders the reliability of the ICQ phase;
- the choice of an appropriate operative volume to avoid a too crowded filter.

7.2.4 IQC

The Identification Quantification and Characterization phase begins with the image acquisition. Indeed, to automate the process as much as possible, to guarantee a lower relevance of the operator and to lower the time requirements for the analysis, the particles to be analysed are selected on the filter through image analysis, selecting a threshold either by an algorithm or by the operator. The first choice to make was the magnification to use: this affects the resolution of the image, the stability of the focus, the opportunity to use the autofocus, the time of the image acquisition. It has to be noted the relevance of the use of proper focus distance to obtain quality spectra. The presence of particles of significantly different dimension from average is to be avoided as they can both confuse the image analysis algorithm and anyway end up out of focus. Another aspect often overseen is the possible movement of the filter on stage as that of the particles on the filter; these aspects gain relevance with the decrease of the particle size analysed and might define the lowest particle size analytical limit, as the stitching error for the tiles (pictures taken by the microscope) that composes the mosaic of the whole image to be analysed.

Once the image is acquired and the coordinates of the particle of interest defined, the laser is directed on those. The parameters of relevance are: laser wavelength, laser power, time of acquisition and repetition of acquisitions. It is well known that the main issue of the Raman technique is the high risk of fluorescence, which can be slightly decreased by the use of different lasers. However, the time requirements linked to this option, make it not suitable for routine MPs analysis. Therefore, one wavelength has to be chosen, and the other parameters decided as best.

The acquired spectra are then compared to reference libraries, and the outcome of the identification depends both on the library used (the reference spectra present in terms of pure polymers, blends, pristine and aged MPs) and on the comparison method used. The method "correlation" is the mostly used in literature, with an identification threshold of 60–70%.

7.2.5 QA/QC

Due to MPs ubiquity, it is of utmost relevance to carry on negative controls to check for contamination. It is also of utmost relevance to carry on "positive controls" as recovery rates with reference material. If the numbers reported as results for MPs content are corrected by blanks results, the way and reasoning behind it should be carefully described, considering the full MPs spectrum: size, polymer, shape.

Blanks were evaluated with ultrapure water type 1 from Arius VF pro by Sartorius. Recovery rates were initially evaluated thanks to the reference material provided by Quasimeme and using green Polyethylene microspheres acquired from Cospheric with size >75 μm.

7.3 Results and Discussion

7.3.1 Pre-treatments

The tested pre-treatments to analyse particles >30 μm (using hence a 10x magnification) showed all comparable results. The pre-treatment choice is hence dictated by practical observations. The use of AA was hindered by the sporadic formation of a film on the 1 μm filter due to the interaction between AA and substances present in water or possibly with the ethanol used to speed up the filtration. The use of AA is hence still deemed of potential interest, but was discarded due to practical aspect to be yet clarified. EDTA also showed a similar effect as HCl in reducing the particles on the filter; however, it might also interact with the ethanol used to speed up the filtration, precipitating on the filter therefore loosing efficiency. The beneficial effect on the particles present on the filters thanks to the use of HCl at a pH of 3 for 15 min is shown in Fig. 7.1.

Regarding the choice of pre-treatments for MPs <20 μm, preliminary trials showed better efficiencies based on duplicates for HCl compared to EDTA, with the latter showing about 25% particles more than the former. The particles were counted by the automatic algorithm implemented in the image analysis software of

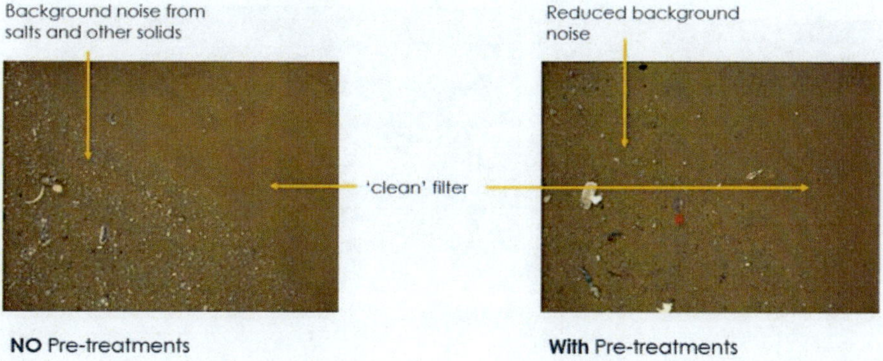

Fig. 7.1 Effect of pre-treatments by HCl

the DXR3, with a 50% magnification, on 25 windows of size 4*FOV (Field of View), chosen randomly on the filter by an algorithm designed on the software R studio. The choice of the windows is dictated by the time available for the analysis. However, these data might be hindered by the high presence of small particles that were later on discovered being released from the glassware.

Therefore, also for lower particles size, the pre-treatment of choice is HCl, although trials are ongoing to test the effect of such pretreatment on the MPs of smaller size range.

7.3.2 *Filtration*

Regarding the filtration phase, the time was significantly reduced by filtering a small amount of ethanol before the sample. This procedure was instead not required when using filters with nominal porosity of 5 μm.

The presence of bubbles was reduced by mixing or gently shaking the sample before filtration, allowing the extra gas produced during the pre-treatments to be released. Bubbles were instead not formed when using the 5 μm filters. The optimal volume to be filtrated on one filter in the present case was observed to be not exceeding 500 mL, to avoid filter overcrowding.

7.3.3 *IQC*

It was noted how the automatic algorithm implemented in the software got confused in the presence of a clean filter of 1μm (mostly when using the 10x magnification), or in the presence of relatively large particles. Therefore, it was decided to use an operator selected threshold, based on two factors: the position of the inflection

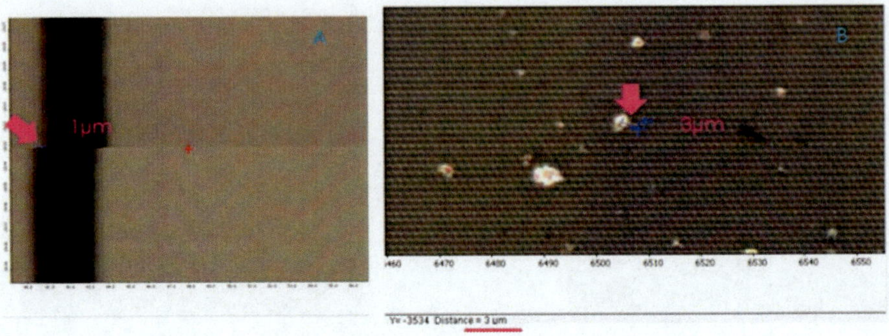

Fig. 7.2 **a** Stitching error, **b** relative movement of particle. 50x magnification

point of the image intensity histogram (image intensity is the parameter used by the algorithm for thresholding) and the image intensity value that allows the non-selection of background filter points. The comparison of particle numbers was in this case highly hindered, but it allowed for a consistent analysis methodology. The algorithm performance is instead consistent in the case of 5µm filters, thanks to the higher contrast between particles and background.

The relative movement of the particles on the filter and the stitching error linked to the filter stage minimum step size, did not allow for the automatic analysis of MPs under 5 µm. This obstacle can be overcome by using a very precise motorized stage (Fig. 7.2). The analysis was anyway possible with an "point and shoot" approach, where the point of interest was chosen one by one manually by the operator. However, this procedure may be considered acceptable for research purposes, not for routine goals.

An acceptable focus must be set on average for all the particles analysed. The use of autofocus is not recommended due to its sensibility to the presence of slightly out of size particles and the time required to acquire the image.

The parameters for proper spectra acquisition are to be set as average for routine goals, whereas the chance of using optimizing tools based on the Signal to Noise ratio can be of high advantage for research purposes. The performance of an average set of parameters compared to optimal ones for each particle is still to be assessed.

7.3.4 QA/QC

Blank results showed random presence of Polyethylene and Polypropylene particles, in an amount comparable to that of the preliminary results obtained with drinking water; therefore, for the presentation of the results to come, it was decided not to apply blank correction.

Regarding recovery rates, the results initially obtained with material acquired by Quasimeme were not considered satisfying, being the recovery rates lower than 60%;

therefore, a rinsing stage after filtration was included. More trials with Cospheric microspheres are ongoing.

7.4 Conclusions

MPs analysis by μ-Raman proved to be challenging due to the many sources of errors and uncertainties that must be necessarily addressed in establishing the optimal protocol. Some errors might be reduced by carefully improving the practical aspects of the analysis; however, it is of utmost relevance to quantify the uncertainties of the number given as a result of the analysis, coupled with a complete description of the protocol used. In this context, it is highly recommended to take as a reference the proposed check index by Cowger et al. [2].

Considering the issues reported in this report, for routine monitoring, a set of three different protocols have been conceived and are going to be developed:

(1) 10X magnification, screening routine (>30 μm).
(2) 20X magnification, specialistic (10–30 μm).
(3) 50X magnification, highly specialistic (5–10 μm).

For practical reasons, the use of 5 μm silicon filters appears more convenient considering the need of an automated analysis for routine monitoring, as the use of HCl pre-treatments. Next research activities will include the evaluation of the effect of the chosen pretreatment on MPs integrity for smaller size ranges, and the screening of DWTPs with the developed method.

References

1. Novotna K, Cermakova L, Pivokonska L, Cajthaml T, Pivokonsky M (2019) Microplastics in drinking water treatment—current knowledge and research needs. Sci Total Environ 667:730–740. https://doi.org/10.1016/j.scitotenv.2019.02.431
2. Cowger W et al (2020) Reporting guidelines to increase the reproducibility and comparability of research on microplastics 74(9)

Chapter 8
Stability of Microplastics: Detection of Potentially Harmful Compounds Produced by Microplastics During Accelerated Aging Studies

Jacopo La Nasa, Greta Biale, Marco Mattonai, Ilaria Degano, Andrea Corti, Valter Castelvetro, and Francesca Modugno

8.1 Introduction

Microplastics (MPs) have now been detected in all the main environmental compartments, and the associated environmental and health hazards have been the focus of intense social, scientific, and media attention [1–3]. Plastic debris should be considered by no means as chemically stable persistent pollutants, but rather as reactive materials. Indeed, synthetic polymers exposed to the environment undergo chemical and physical degradation processes leading not only to mechanical but also molecular fragmentation [4–6]. In this work we tested different analytical approaches based on pyrolysis, gas chromatography, and mass spectrometry to evaluate the potentially harmful molecular fragments generated by MPs during accelerated ageing.

In this study samples of micronized reference polymers (high-density and low-density polyethylene, polypropylene, and polystyrene) were photoaged in SolarBox and treated with specifically optimized extraction protocols to characterize the degraded soluble fractions produced during irradiation. The analyses were performed both on the solvent-soluble fraction of the polymers and the volatile organic compounds produced and released by the reference MPs [7].

Accelerated photo-oxidative aging of the same four reference MPs was also performed directly in artificial seawater to evaluate the production of water-soluble chemical species [8].

Together with the evaluation of the production of soluble species from reference polymers, a selection of environmental micro- and macro-plastics were also analyzed

J. La Nasa (✉) · G. Biale · M. Mattonai · I. Degano · A. Corti · V. Castelvetro · F. Modugno
Department of Chemistry and Industrial Chemistry, University of Pisa, Pisa, Italy
e-mail: jacopo.lanasa@unipi.it

M. Cocca et al. (eds.), *Proceedings of the 3rd International Conference on Microplastic Pollution in the Mediterranean Sea*, Springer Water,
https://doi.org/10.1007/978-3-031-34455-8_8

to characterize and quantified the emission of volatile organic compounds (VOCs) [9].

All the analysis were performed by combining different analytical approaches based on mass spectrometry: analytical pyrolysis coupled with gas chromatography and mass spectrometry (Py-GC–MS) was used to study the polymers and solvent extracts, GC/MS was applied for the study of the water-soluble low molecular weight species after extraction, and selected ion flow tube mass spectrometry (SIFT-MS) was applied for the study of the VOCs.

The combination of artificial ageing and extraction procedures, together with the multi-analytical approach, provided an initial but comprehensive picture of the chemical nature of degradation products released by different polymers present as microplastics in the environment, significantly contributing to our understanding of the fate of this kind of pollutants.

8.2 Experimental

8.2.1 Reference Materials and Environmental Samples

The artificial ageing studies were performed using micronized polypropylene (PP), polystyrene (PS), low-density polyethylene (LDPE) and high-density polyethylene (HDPE), with particle size in the 857–509 μm range. The complete features of the reference microplastics are reported in [7].

The study of the VOCs emission was performed on 39 macro- and microplastic samples collected as sieve fractions (2 mm) from sand samples collected in a marine beach in northern Tuscany (Marina di Vecchiano, Pisa, Italy). Detailed information on the samples is available in [9].

8.2.2 Artificial Ageing of Reference Microplastics

8.2.2.1 Artificial Ageing for the Study of the Solvent-Soluble Fraction

Micronized PP, PS, PET, LDPE and HDPE were artificially aged for four weeks using a solar-box system (CO.FO.ME.GRA. Srl, Milan, Italy) equipped with a Xenon-arc lamp and outdoor filter. The conditions for the aging were: temperature 40 °C, irradiance 750 W/m^2, relative humidity around 60% [7].

8.2.2.2 Artificial Ageing for the Study of the Water-Soluble Fraction

Five mL of artificial seawater were added to quartz tubes, each containing 1.5 mg of the reference polymer. The quartz tubes with the polymers in artificial seawater were placed in the SolarBox equipped with a Xenon lamp and outdoor UV filter, operating at T $\approx$ 40 °C and 750 W/m^2 irradiance, and aged for two weeks [8].

8.2.3 Pyrolysis-Gas Chromatography-Mass Spectrometry Conditions (Py-GC-MS)

The analyses were performed using a multi-shot pyrolyzer EGA/PY-3030D (Frontier Lab, Japan) coupled with an 8890 gas chromatograph, combined with a 5977B mass selective single quadrupole mass spectrometer detector (Agilent Technologies, US). The complete conditions used for the studies are reported in [7].

8.2.4 Gas Chromatography-Mass Spectrometry (GC–MS)

The GC–MS instrumentation consisted of a 6890 N gas chromatograph coupled with a 5975 mass selective detector single-quadrupole mass spectrometer. The complete instrumental conditions and the semi-quantitative parameters are reported in [8].

8.2.5 Selected Ion Flow Tube-Mass Spectrometry (SIFT-MS)

The environmental samples were sealed in 22 mL gas-tight glass vials and kept at 60 °C for 20 min equilibration time before the analysis. The analyses were performed using a Voice 200ultra instrument (SYFT Technologies, New Zealand). The reagent ions used for ionizing the analytes were H_3O^+, NO^+, and O_2^+. The complete instrumental conditions are reported in [9].

8.2.6 Infrared Spectroscopy Analysis

Infrared spectra in the mid-IR region (700–4000 cm^{-1}) were recorded with a Perkin Elmer Spectrum Autoimage System microscope equipped with an Attenuated Total Reflectance (ATR) module with germanium crystal; each spectrum is the result of 64 scans accumulation at 4 cm^{-1} spectral resolution. The lateral spatial resolution corresponds to the contact area with the germanium crystal tip (30–40 micron) [9].

8.3 Results and Discussion

8.3.1 *Evaluation of the Solvent-Soluble Fraction of Microplastics*

To study the formation of solvent-soluble chemical species during the artificial ageing of microplastics we performed a Soxhlet extraction using dichloromethane for polyolefins and methanol for polystyrene. These solvents were chosen to avoid dissolution of the polymeric bulk. With this approach we selectively removed from the polymer only the low molecular weight fraction. Both the solvent extracts and the insoluble residues were then analyzed by Py-GC-MS. For solvent extracts, in situ derivatization was also performed by adding hexamethyldisilazane (HMDS) to the sample immediately before analysis, to improve the chromatographic performances towards compounds with polar functional groups. Figure 8.1 reports the flowchart of the procedure for the investigation of the polymer degradation products generated upon accelerated photo-oxidative aging.

This study highlighted that the stability towards the ageing of polyolefins was related to their structure, in particular on the occurrence of tertiary carbon atoms in the polymeric chain, with consequent higher stability of the radicals resulting from the dehydrogenation. For this reason, LDPE was more prone to the degradation in respect to HDPE, while PP after the ageing was more degraded then the two polyethylenes. The oxidation products of HDPE and LDPE were mainly characterized by the presence of long chain alcohols, aldehydes, ketones, carboxylic acids, and hydroxy acids (chain lengths up to 35 carbon atoms). The soluble fraction of PP was mainly characterized by the presence of short chain mono- and dicarboxylic acids.

PS was significantly degraded by artificial photo-oxidative aging. In detail the methanol soluble fraction was characterized by the presence of alcohols and carboxylic acid, the most abundant being benzoic and 1,4-benzenedicarboxylic acids.

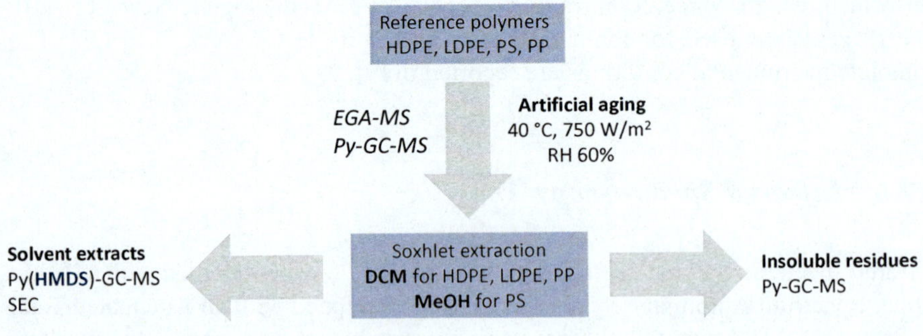

Fig. 8.1 Sample treatment for Py-GC-MS analysis of artificially aged plastic particles [7]

Additional SEC analysis highlighted the presence of chemical species with molecular weights in the range 500–10,200 Da [7].

8.3.2 Evaluation of the Water-Soluble Fraction of Microplastics

To study the formation of water-soluble oxidation products during photo-oxidative artificial ageing the micronized polymers were directly aged in synthetic sea water. After two weeks of ageing, the mix was filtered in order to remove the insoluble residues. The water extract was subsequently extracted with hexane, and diethyl ether after acidification with hydrochloric acid. The two extracts were combined in order to analyze them in a single chromatographic run, evaporated to dryness under nitrogen stream, and subjected to silanization with N,O-bistrimethylsilyltrifluoroacetamide (BSTFA) before GC-MS analysis. Figure 8.2 reports the flowchart of the procedure for the investigation of the water-soluble polymer degradation products generated upon accelerated photo-oxidative aging in artificial sea water.

The extract of LDPE was characterized by the presence of linear hydroxy acids in the C_2–C_{12} range, and a broader range of C_8–C_{22} monocarboxylic and C_3–C_{17} dicarboxylic acids. The total mass of soluble species from LDPE was 7.5 ppm. The extract of PP was characterized by the presence of branched dicarboxylic acids in the C_4–C_{14} range and branched monocarboxylic acids up to C_{14}, with 2,4-dimethyl-hexanedioic acid as the most distinctive and abundant. A total of 3.0 ppm of leached oxidation products was calculated for PP.

The extract of PS was characterized by benzoic acid, phenylacetic acid, 4-hydroxybenzaldehyde, hydroxy-benzoic acids, as main degradation products, along with several variously oxidized styrene dimers and trimers. A total of 35.6 ppm of

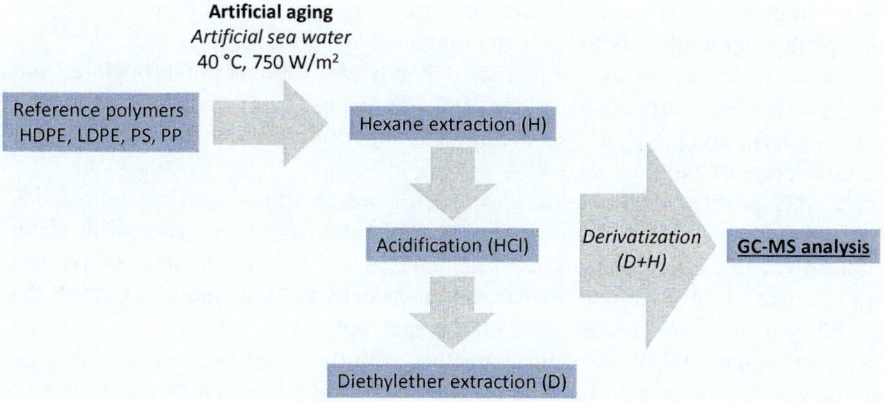

Fig. 8.2 Sample treatment for GC_MS analysis of chemical species leached from plastic particles [8]

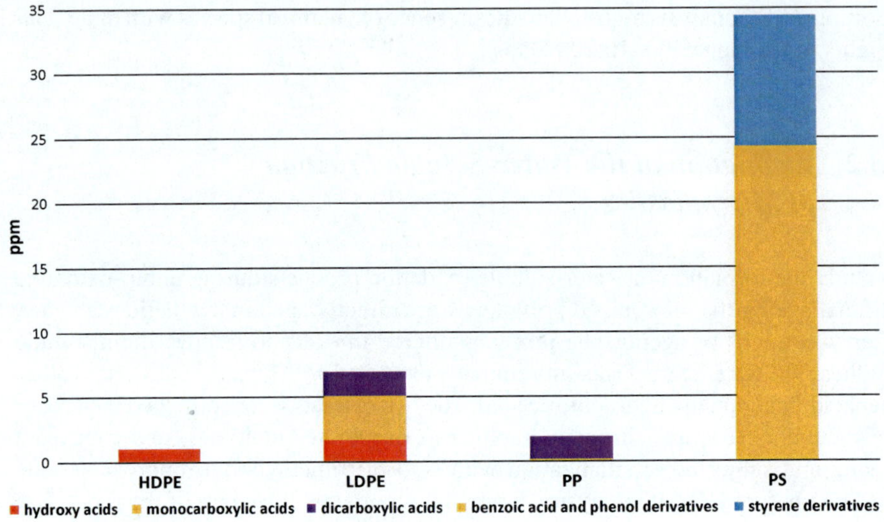

Fig. 8.3 Amounts of all the main categories of compounds detected in the leachates of polyolefins and polystyrene [8]

leached oxidation products was calculated for PS. Figure 8.3 reports the amounts of all the main categories of compounds detected in the leachates of polyolefins and polystyrene.

8.3.3 Study of the VOCs Emission from Environmental Microplastics

The environmental plastic debris were preliminary characterized by infrared spectroscopy in order to identify the type of polymer: 54% of the samples were identified by infrared spectroscopy as low-density polyethylene, 38% as polypropylene, and 8% as polyethylene terephthalate. The SIFT-MS acquisitions were performed using $[H_3O]^+$, $[NO]^+$, and $[O_2]^+$ as reagent ions. The experimental flowchart for the VOCs analysis is reported in Fig. 8.4.

The VOC profiles of polyolefins plastic debris were characterized by the presence of the ions characteristic of aldehydes, alcohols and carboxylic acids with chain lengths up to 12 carbon atoms. The VOCs emitted by PET were mainly compounds with benzene, styrene, styrene oxide, and benzaldehyde. PCA analysis showed the possibility to discriminate the type of polymer only at high degree of oxidation (Fig. 8.4). Finally, the PCA results, together with the quantitative results showed that the samples from the winter berm, characterized by longer accumulation time in respect to the foreshore and the summer berm, were those characterized higher oxidation degrees, and thus by richer VOC profiles.

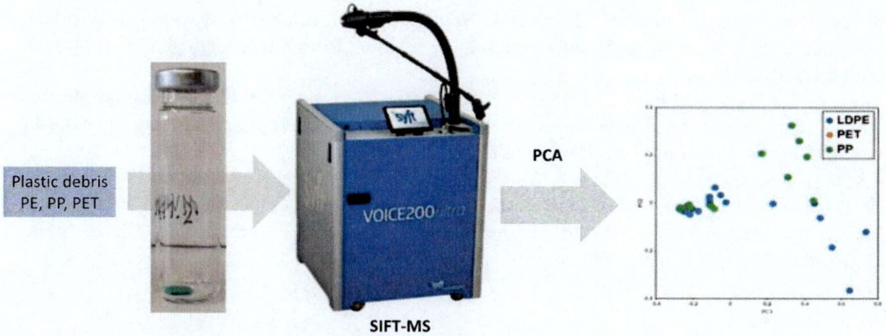

Fig. 8.4 Experimental flowchart for the SIFT-MS analysis of VOCs [9]

8.4 Conclusions

Microplastics are dynamic systems that undergo physical–chemical changes during natural weathering, and release different classes of oxidation products in the environment.

Artificial ageing of microplastics is crucial to simulate their behaviour in the environment and to evaluate the real long-term impact of these pollutants into the ecosystem. Selective Py-GC-MS and GC-MS analysis of degradation products associated to MPs contribute to a complete understanding of the risk associated to this type of pollutants.

References

1. Goßmann I, Süßmuth R, Scholz-Böttcher BM (2022) Plastic in the air?!—spider webs as spatial and temporal mirror for microplastics including tire wear particles in urban air. Sci Total Environ 832:155008
2. Kirstein IV, Hensel F, Gomiero A, Iordachescu L, Vianello A, Wittgren HB, Vollertsen J (2021) Drinking plastics?—quantification and qualification of microplastics in drinking water distribution systems by μFTIR and Py-GCMS. Water Res 188:116519
3. Fabbri D, Tartari D, Trombini C (2000) Analysis of poly(vinyl chloride) and other polymers in sediments and suspended matter of a coastal lagoon by pyrolysis-gas chromatography-mass spectrometry. Anal Chim Acta 413:3–11
4. Toapanta T, Okoffo ED, Ede S, O'Brien S, Burrows SD, Ribeiro F, Gallen M, Colwell J, Whittaker AK, Kaserzon S, Thomas KV (2021) Influence of surface oxidation on the quantification of polypropylene microplastics by pyrolysis gas chromatography mass spectrometry. Sci Total Environ 796:148835
5. De Falco F, Avolio R, Errico ME, Di Pace E, Avella M, Cocca M, Gentile G (2021) Comparison of biodegradable polyesters degradation behavior in sand. J Hazard Mater 416:126231
6. Bianco A, Sordello F, Ehn M, Vione D, Passananti M (2020) Degradation of nanoplastics in the environment: reactivity and impact on atmospheric and surface waters. Sci Total Environ 742:140413

7. Biale G, La Nasa J, Mattonai M, Corti A, Vinciguerra V, Castelvetro V, Modugno F (2021) A systematic study on the degradation products generated from artificially aged microplastics. Polymers 13:1997
8. Biale G, La Nasa J, Mattonai M, Corti A, Castelvetro V, Modugno F (2022) Seeping plastics: potentially harmful molecular fragments leached out from microplastics in seawater during accelerated ageing. Water Res 219:118521
9. La Nasa J, Lomonaco T, Manco E, Ceccarini A, Fuoco R, Corti A, Modugno F, Castelvetro V, Degano I (2021) Plastic breeze: volatile organic compounds (VOCs) emitted by degrading macro- and microplastics analyzed by selected ion flow-tube mass spectrometry. Chemosphere 270:28612

Chapter 9
Plastic Recycling and Microplastic Emissions

Michael Süß, Christian Marschik, and Jörg Fischer

9.1 Introduction

Plastic waste and its detrimental impact on the environment is an increasing concern of human society. Furthermore, the growing demand of plastic products leads to a rise of needed raw materials such as crude oil, ascending CO_2 footprint and more waste. Currently, most of the plastic waste worldwide is going to be discarded. Hence, the linear economy of plastics is overpronounced. However, plastics recycling, a way to enter the circular economy of plastics, enables a more sustainable future for plastic products and reduces the CO_2 footprint of the plastic industry. In 2020 the worldwide and European (EU27 + NO + CH + UK) plastic production was around 367 and 55 Mt, respectively [1]. In Europe (EU + NO + CH + UK) 29.5 Mt of plastic post-consumer waste was collected and 34.6, 23.4 and 42% were recycled, landfilled, and sent to energy recovery, respectively [1]. These numbers suggest the possibility to increase recycling rates and consequently reduce the need of virgin material for new plastic products. Despite all the associated benefits due to recycling, the recycling process also emits waste- and or side streams.

During the mechanical recycling of post-consumer plastic waste, the collected material is often pre-treated by shredding, sorting, and washing with water. Hence, process water accrues, which can to a certain extent be kept in the process, however, might be contaminated and released. One source of contamination could be microplastics (MPs) which are formed due to friction and temperature during processing. These MP particles with dimensions smaller than 5 mm according to the Environment Agency Austria [2] potentially could be emitted to the environment.

M. Süß (✉) · C. Marschik
Competence Center CHASE GmbH, Linz, Austria
e-mail: michael.suess@chasecenter.at

J. Fischer
Institute of Polymeric Materials and Testing, Johannes Kepler University Linz, Linz, Austria

M. Cocca et al. (eds.), *Proceedings of the 3rd International Conference on Microplastic Pollution in the Mediterranean Sea*, Springer Water,
https://doi.org/10.1007/978-3-031-34455-8_9

However, to the best of our knowledge the occurrence of MPs during a recycling process is scarcely investigated. Suzuki et al. investigated the MP generation during mechanical recycling of electronic plastic waste (EPW), polyethylene terephthalate (PET) bottle waste, and household plastic waste [3]. In this study 5.8, 0.13 and 0.014 to 5.8 t/year, respectively, of microplastic were estimated to be discharged.

Therefore, in this ongoing study the occurrence of MP during the pre-treatment of post-consumer polypropylene (PP) is investigated to generally identify sources of MP during the washing and shed more light on possible microplastic emissions. In addition, a cost effective and easier way to quantify MP is evaluated, followed by spectroscopical classification with Raman spectroscopy.

9.2 Experimental

9.2.1 Materials

Washed post-consumer PP flakes were used as a sample for evaluating the applicability and accuracy of measuring the Total Suspended Solids (TSS) for microplastic analysis as well as for washing trials.

9.2.2 Methods

In the first part, the applicability in terms of MP analysis and its accuracy of measuring the TSS, washed and shredded post-consumer polypropylene flakes were cryogenically milled. Weighted samples were transferred into a 1 L volumetric flask and filled up with deionized water. Subsequently filtration was conducted on microporous silicon membrane with a pore diameter of 10 μm and an effective filter area of 13 mm.

In the second part, it is verified if MP are emitted by pre-treated polypropylene flakes. For this purpose 0.8 kg of 4–8 mm sized flakes were washed in an industrial washing machine (Miele Professional PW818) to simulate an industrial cleaning process. The effect of temperature was started to be investigated and fleet ratio and PP loading will be additionally examined. The washing water of the unloaded washer is considered as reference.

After washing, an aliquot sample of the effluent was filtered, dried and consequently spectroscopically analysed.

9.2.2.1 Analytical Techniques

Filtered samples were dried at 50 °C in a drying oven (Heraeus Type T50 0.7) and the weight was measured using a scale (Satorius QUINTIX224). Scanning for possible MP particles was conducted using a Raman Spectroscope. Raman Imaging was conducted at 5 μm increments with a wavelength of 532 nm, laser power of 15 mW and an integration time of 0.25 s.

9.3 Results and Discussion

9.3.1 Results

Results of the first part, the evaluation of applicability and accuracy of the TSS measurements to quantify MP contamination during washing of post-consumer PP is shown in Fig. 9.1. At least triplicates were measured, and the error bars indicate the determined maximum and minimum weight of a sample. As shown at a weight fraction >5.5 mg the evaluation did not underestimate the value with three measurements. This will be further investigated.

After quantification of the samples Raman spectroscopy was conducted to classify MP residue on the filter. In Fig. 9.2, a microscopic image of a sample is depicted, showing an about 1 × 1 mm window indicated in red. The generated spectra within this window are showing PP and the used silicon filter (component 2). Based on these spectra a Raman Image is created.

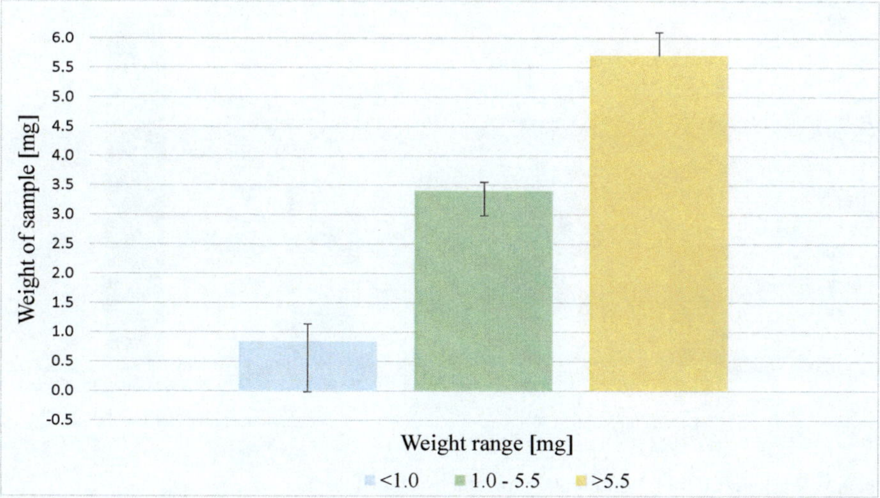

Fig. 9.1 Weight of three pre-defined weight-fractions for microplastic

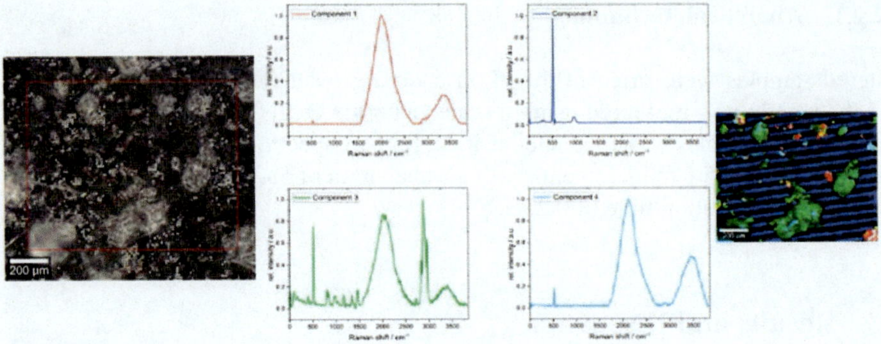

Fig. 9.2 Left: Microscopic image; middle: Raman spectra (Component 1, 3 and 4 representing PP; Component 2 representing silicon filter) and right: Raman image (overlaying of Raman spectra) of dried filter

As the evaluation of applicability and accuracy of the TSS measurements and subsequent spectroscopic analysis is in an advanced stage, especially the TSS evaluation, first washing trials (the second part of this study) with post-consumer PP were conducted.

The washing trial, depicted in Fig. 9.3, is showing the MP generation in dependence of the temperature. As of now two different temperatures were tested, 20 and 90 °C as low- and high temperature, respectively.

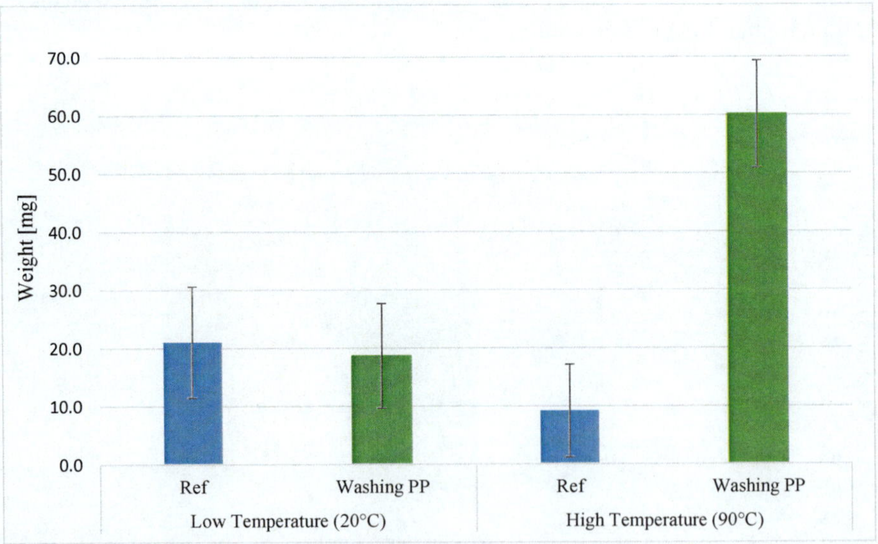

Fig. 9.3 Washing trials of 2 different temperatures (20 and 90 °C)

9.3.2 Discussion

Presented results for TSS measurements (first part of the study) show an underestimation of sample weight at weight fractions of <1.0 and 1.0–5.5 mg which are −0.85 and −0.42 mg, respectively. The effect gets more pronounced as sample-weight-fraction decline. The underestimate potentially causes undervalued release of MP, however in comparison to a potential overestimate the possible overvalue are significantly lower. This gathered knowledge enables a more accurate prediction of emitted MP. The measurement of TSS is possible to a certain extent e.g., treating post-consumer waste might need an additional step to remove organics adhering to the surface. Furthermore, Raman spectroscopy of MP after filtration was tested and showed a moderate expenditure of time and distinctive classification between the used PP and silicon filter.

The first washing trial, depicted in Fig. 9.3, at a temperature of 20 °C shows an overlapping standard deviation between the reference (empty washer) and washing PP and hence no distinctive differentiation in terms of MP generation is possible. Since Raman analysis is not completed at this point it is supposed that the weight of the reference is based on particles from the washer or piping. It is also assumed that when washing PP at 20 °C no significant amount of MP is released, and the measured weight is based on particles originating from the washer or piping. However, at a maximum temperature of 90 °C a significant difference between the reference and washing PP can be seen. Considering the standard deviation, a minimum of 23.2 mg and maximum of 59.7 mg of potentially MP could be release by washing post-consumer PP waste at 90 °C. To enhance the accuracy in terms of minimizing the standard deviation the whole washing water should be filtered and analysed. Analysing an aliquot sample might not be sufficient, since the low density of PP does not enable a complete homogenisation of the whole washing effluent. Further factors such as fleet ratio of PP and washing water need to be tested, verifying the effect on MP emissions. At this point, it needs to be emphasised that the total amount of emitted MP possibly changes but not the result that PP MP is emitted.

9.4 Conclusions

Presented results are revealing the generation of MP emissions during washing of post-consumer PP flakes. Deploying the measurement of TSS to quantify MP emissions in washing effluents are possible to a certain extend. In addition, it is an easy and cost-effective possibility. Quantification needs to be enhanced in terms of analysing the whole washing water instead of an aliquot sample. Further trials need to be conducted to verify if other factors such as fleet ratio of PP and washing water influence MP emissions.

Acknowledgements The authors acknowledge financial support through the COMET Centre CHASE, funded within the COMET—Competence Centers for Excellent Technologies programme by the BMK, the BMDW and the Federal Provinces of Upper Austria and Vienna. The COMET programme is managed by the Austrian Research Promotion Agency (FFG).

References

1. Plastics Europe (2021) Plastics the fact 2021, Plast. Eur. Mark. Res. Gr. Conversio Mark. Strateg. GmbH., pp 1–34. https://plasticseurope.org/
2. Liebmann B, Sexlinger K (2019) Mikroplastik in der Umwelt - Statusbericht
3. Suzuki G, Uchida N, Tuyen LH, Tanaka K, Matsukami H, Kunisue T, Takahashi S, Viet PH, Kuramochi H, Osako M (2022) Mechanical recycling of plastic waste as a point source of microplastic pollution. Environ Pollut 303:119114. https://doi.org/10.1016/j.envpol.2022.119114

Chapter 10
Additives, Plasticizers, and Small Microplastics (5–100 μm) in Wet and Dry Depositions from an Urban Area of Venice, Italy

Beatrice Rosso, Carlo Barbante, Andrea Gambaro, and Fabiana Corami

10.1 Introduction

The atmosphere is considered one of the most important pathways for the transport of microplastics (MPs) from urban areas to far and remote places [1, 2].

Atmospheric depositions consist of dry and wet depositions, and they may be a positive driver in atmospheric pathways of emerging pollutants. Dry and Wet depositions are primary factors influencing the fallout flux as they contribute to scavenging processes for aerosol particles, including MPs.

However, deposition of individual rain/snowfall events has not been widely studied in scientific literature, and it is essential to investigate the atmospheric pathways and sources of MPs. The concentration of airborne microplastics is most likely associated with anthropogenic activities, population density, industrialization levels, and meteorological conditions such as rainfall, snowfall, and wind. Due to their smaller sizes, the smallest particles (SMPs 5–100 μm) in the air are overlooked in scientific literature, and they could be transported over long distances through atmospheric currents and be inhaled by biota and humans [3–5].

Plastic objects are a mix of plastic polymers, additives, and plasticizers; when plastic objects get fragmented, additives and plasticizers can be released into the

B. Rosso (✉) · C. Barbante · A. Gambaro · F. Corami
Department of Environmental Sciences, Informatics, and Statistics, Ca' Foscari University of Venice, Via Torino, 155, 30172 Venezia-Mestre, Italy
e-mail: Beatrice.rosso@unive.it

C. Barbante · F. Corami
Institute of Polar Sciences, National Research Council, Venezia-Mestre, Italy

M. Cocca et al. (eds.), *Proceedings of the 3rd International Conference on Microplastic Pollution in the Mediterranean Sea*, Springer Water,
https://doi.org/10.1007/978-3-031-34455-8_10

environment. Besides, these compounds can be transported through different environmental compartments contributing to potential toxic effects on different organisms [6].

The different experimental methods and techniques used can affect the comparison and integration of results from different studies presented on MPs. Standardized operating protocols have not yet been established for sample collection and subsequent analysis for their quantification and chemical identification of SMPs, plasticizers, and additives in the atmospheric compartment [3]. Further, there is an urgent need for data on the exposure risk of wet and dry depositions that involve SMPs fluxes in terrestrial and aquatic ecosystems [3].

SMPs, additives, plasticizers, and natural and non-plastic synthetic fibers (APFs) were investigated in wet and dry depositions from an urban area of Mestre (Venice-Italy). A pre-treatment method was developed for the quantification and chemical identification of SMPs and APFs using a Micro-FTIR to avoid further degradation/denaturation of these compounds [7]. From the results, the abundance SMPs and AFPs observed in a wet deposition is higher than in dry periods due to the scavenging effect of rain. Different sources from the urban area could affect the abundance and distribution of SMPs and APFs. Different additives and plasticizers investigated could be a good proxy of SMPs' presence in the environment, for instance, road dust resuspension.

10.2 Experimental

10.2.1 Sampling and Study Area

The research was performed in Mestre city, near Venice, in the north of Italy. The sampling site is located on the roof of the scientific campus of the Ca' Foscari University of Venice.

The dry deposition was collected at the end of the 2 weeks from December 2020 to March 2021; wet depositions samples were collected after each rainfall event. A wet-dry automatic collector (Wet&Dry Sampler FAS005AB, MTX, Padova, Italy) was equipped with two metal vessels and a rain sensor to collect wet or dry depositions in concomitance with the precipitation, respectively. Wet depositions were recovered in a previously decontaminated Erlenmeyer flask with Ultrapure Water (UW; Elga Lab Water, Veolia, High Wycombe, UK) and a 50% Ethanol (205 Superpurity Solvent >99.9%, Romil, Cambridge UK)/Methanol (absolute, for HPLC, ≥99.8%, Sigma Aldrich, Merck, Darmstadt Germany) solution (volume ratio 1:1:1 respectively). Dry depositions were carefully recovered in a previously decontaminated Erlenmeyer flask with 300 ml of ethanol/methanol solution and 300 ml of UW. All flasks were then stored at 4 °C until the pre-treatment.

10.2.2 Methods

10.2.2.1 Preparation of Samples

Each flush was pretreated using *soft digestion* with hydrogen peroxide 30%, microwave-digestion (Ethos1-Milestone), and stirring on a multipurpose orbital shaker. Then, the filtrations of each sample were performed according to Corami et al. [7]. The filtration was performed under a fume hood in a plastic-free Clean Room ISO 7, using a glass vacuum filtration system (VWR International, Milan, Italy) and aluminum oxide filters (0.2 μm ANODISC (Anopore Inorganic Membrane), Whatman, Merck Darmstadt Germany). Specifically, a solution of ethanol-methanol and ethanol was added initially to cleanse the filter; then the aliquot of wet and dry depositions samples was flushed several times alternatively with a mixture of ethanol-methanol and ethanol. Finally, ethanol and UW were employed to rinse the filters. Reagent and procedural blanks were filtered accordingly. All filters were stored at room temperature for 72 h in Clean Room in previously decontaminated glass Petri dishes. Extraction of each sample was run in duplicate.

10.2.3 QA/QC

For the SMPs and APFs sampling, two glass jars were placed inside the automatic sampler. Then, all samples were carefully covered with aluminum foils and transported to the cleanroom (ISO 7) to be processed. Decontamination and pre-treatment procedures were performed at CNR-ISP Venice in a plastic-free cleanroom ISO 7. This cleanroom is entirely free of plastic materials, even in the air pre-filters, and it consists of a controlled-atmosphere laboratory where atmospheric pressure, humidity, temperature, and particle pollution are controlled. Operators wore cotton lab coats and nitrile gloves to prevent plastic contamination. All Samples were pre-treated (extraction and purification) and filtered in batches on aluminum oxide filters (ANODISC filters, 0.2 μm, 47 mm, Whatman™; Merck, Darmstadt, Germany). The pre-treatment procedures and filtration were performed under a decontaminated steel fume hood. Glassware and steelware were previously washed with a 1% Citranox® solution, rinsed with ultrapure water, and decontaminated with acetone, methanol, and ethanol. Reagents and procedural blanks were performed for each batch. After filtration, all filters were stored in decontaminated glass Petri dishes covered with aluminum foil.

10.2.3.1 Analytical Techniques

A Micro-FTIR Nicolet™ iN™ 10 infrared microscope (Thermo Fisher Scientific, Madison, WI, USA), equipped with an ultra-fast motorized stage and liquid-nitrogen-cooled MCT detector (mercury cadmium telluride detector), was employed for the analysis. The settings were: transmittance mode, a spectral range of 4000–1200 cm^{-1}, 100-μm step size scanning (spatial resolution), 100–100 μm aperture, and 64 co-added scans at a spectral resolution of 4 cm^{-1}.

Microscopic counting was performed according to Corami et al. [7]. In this work, at least 20 known-sized areas (i.e., count fields) were randomly chosen with no overlapping on the surface of the filter and a significant number of particles (250–350 particles per count field) were analyzed using the PARTICLES WIZARD of the Omnic™ Picta™ software. The identification match percentage was ≥65%; when operating with this software section, the optimal range of match percentage is between 65 and 75%. Moreover, particle sizes (length and width) were collected using the Imaging of PARTICLE WIZARDS.

10.3 Results and Discussion

The pre-treatment method optimized in this study allows for concurrent extraction of all SMPs and APFs particles and does not contribute to these particles' further degradation/denaturation. Many pre-treatment methods employ strong oxidizing agents or strong acids, which can modify particle sizes and contribute to discoloration, degradation, and loss of several polymers, especially nylon, using high temperatures. These methods can contribute to the loss of polymers considering their glass transition temperature. Different particles and fibers are observed with the microscope part of micro-FTIR (Fig. 10.1). However, most of the fibers found were Cellulose or rayon, hence the chemical identification of the fibers and particles in real environmental samples is important to discriminate.

The first results of SMPs and APFs presented in the wet and dry depositions are simultaneously quantified (microscopic count) and identified via Micro-FTIR. The most common plastic typologies found in one of the wet and dry depositions samples are shown (Fig. 10.2).

The most abundant polymers were PE HD (Polyethylene high density), PA (Nylon), PARA (Polyphtalamide), and acrylic polymers. Between wet and dry samples, there is a correlation between the different sampling periods. Generally, dry depositions had a lower concentration of SMPs compared to respective wet depositions. Also, different APFs were identified in each sample; the most abundant were cosmetics, surfactant agents and additives, and vulcanizing agents used for tires, probably generated by the wear of tires on the highways and roads and resuspended by traffic vehicles.

Fig. 10.1 One of the count field analyzed via micro-FTIR from a wet sample

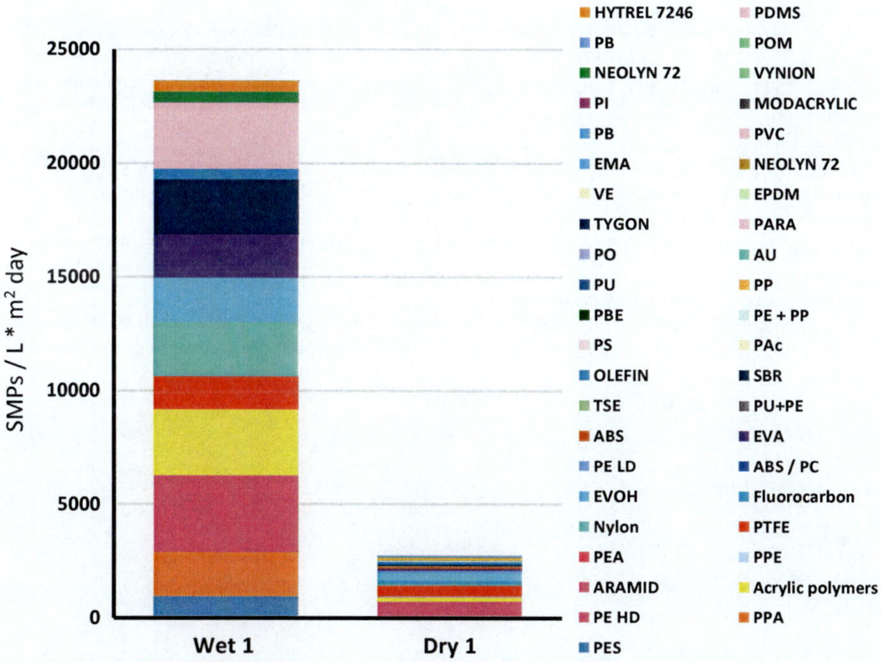

Fig. 10.2 Polymer typologies identified in a wet and in a dry deposition sample

10.4 Conclusions

SMPs and APFs have been detected and quantified in dry and wet deposition in an urban center area of Venice. Development of the pre-treatment procedure was carried out with a strict QA/QC procedure control to avoid any potential plastic contamination. Due to the lack of data on SMPs and APFs in atmospheric compartments in the existing scientific literature, these preliminary results are important to evaluate potential impacts on biota and human health.

References

1. Evangeliou N, Grythe H, Klimont Z (2020) Atmospheric transport is a major pathway of microplastics to remote regions. Nat Commun 11:3381. https://doi.org/10.1038/s41467-020-17201-9
2. Zhang Y, Gao T, Kang S, Sillanpää M (2019) Importance of atmospheric transport for microplastics deposited in remote areas. Environ Pollut 254(Part A):112953. https://doi.org/10.1016/j.envpol.2019.07.121
3. Chen G, Feng Q, Wang J (2020) Mini-review of microplastics in the atmosphere and their risks to humans. Sci Tot Environ 703:135504. https://doi.org/10.1016/j.scitotenv.2019.135504
4. Yuan Z, Li H, Lin L, Pan Y, Liu S, Hou R, Xu X (2022) Occurrence and human exposure risks of atmospheric microplastics: a review. Gondwana Res 108:200–212. https://doi.org/10.1016/j.gr.2022.02.001
5. Luo Z, Wang Z, Yang L, Gao T, Zhang Y (2022) A review of analytical methods and models used in atmospheric microplastic research. Sci Tot Environ 828:54487. https://doi.org/10.1016/j.scitotenv.2022.154487
6. Beiras R, Verdejo E, Campoy-López P, Vidal-Liñán L (2021) Aquatic toxicity of chemically defined microplastics can be explained by functional additives. J Hazard Mater 406:124338. https://doi.org/10.1016/j.jhazmat.2020.124338
7. Corami F, Rosso B, Morabito E, Rensi V, Gambaro A, Barbante C (2021) Small microplastics (<100 μm), plasticizers and additives in seawater and sediments: oleo-extraction, purification, quantification, and polymer characterization using Micro-FTIR. Sci Tot Environ 797:148937. https://doi.org/10.1016/j.scitotenv.2021.148937

Chapter 11
Detection of Microplastics in Marine Sediments: Results from Three Italian Coasts

Valentina Balestra, Federica Trunfio, Cassandra Caione, Evgenia Tsymbaliuk, Paola Marini, and Rossana Bellopede

11.1 Introduction

The presence and dangerousness of microplastics (MPs) in marine environments are globally recognized [1–5]. Their definition in literature and normative is not clear: despite in scientific literature MPs refer to dimension minor then 5 mm, the recent technical report ISO/TR 21960 [6] defines as MP any solid plastic particle insoluble in water with a dimension between 1 μm and 1 mm, while particles between 1 and 5 mm are defined macroplastics.

The MP criticalities are related especially to their small size [2]. Evidence of MP dangerousness for human health have been highlighted for particles smaller than 20 μm, penetrating organs and cell membranes [7].

The main sources of sea water MP pollution are linked to water from treatment plants, rivers contamination, and ships/boats that surf rivers and seas for commercial or trade purposes. The abundance of plastic debris on the beaches could be associated to coastal urban centres and their economic activities, plastic debris displacement related to marine currents [8]. MPs small dimension make them easily transportable, also to long distances, therefore, MP contamination in marine environment could be related to the atmospheric transport too [9, 10].

The content of MPs in marine and beach sediments is very variable, ranging from few hundred of MPs/kg to 12,000 MPs/kg [6].

V. Balestra · F. Trunfio · C. Caione · E. Tsymbaliuk · P. Marini · R. Bellopede (✉)
Department of Environment, Land and Infrastructure Engineering (DIATI), Politecnico di Torino, Torino, Italy
e-mail: rossana.bellopede@polito.it

M. Cocca et al. (eds.), *Proceedings of the 3rd International Conference on Microplastic Pollution in the Mediterranean Sea*, Springer Water,
https://doi.org/10.1007/978-3-031-34455-8_11

Despite plastic has been recognized universally as global contaminant since 2011 [11], methods to determine plastic occurrence reach an increase in scientific production only from 2016. However, at the moment, a standardized procedure for MP sampling, treatment and identification does not exist.

Technical report ISO/TR 21960:2020 [6] gives an overview concerning MP pollution knowledge and recognized methodologies for sampling, analysis and identification of plastic contaminants.

On a regulatory level, the European Union and the Framework Directive on the Strategy for the Marine Environment [12] (Directive 2008/56/EC) promotes the adoption of complex strategies for prevention, protection and conservation of the marine environment related to pressures and environmental impacts of anthropogenic activities.

Recently, international working groups were born, such as "PRIORITY—Plastics Monitoring Detection Remediation Recovery" COST Action [13] (CA20101), focused on developing, implementing, and consolidating strategies to tackle the global challenges of micro- and nano-plastics in the environment.

The aim of this work is to investigate and discuss the results of MP pollution monitoring (particles below 5 mm) in sediments of three Italian beaches, in order to test a reliable method to detect MP in sediments. A focus on MPs definitions and classification was made, as well as records on sampling and treatment procedures.

11.2 Experimental

This research deals with the separation and identification of MP particles present in beach sand sampled in three different Italian coasts (Fig. 11.1): Imperia (Liguria), Metaponto (Basilicata) and Villa San Giovanni (Calabria).

11.2.1 Materials

In Fig. 11.1 the three examined Italian beaches are shown. The three beaches are related to very different economic activities. For each site at least 400 g of sediment were collected from different sampling areas, chosen at different distances from the sea.

11.2.2 Methods

All samples were dried at 40 °C to constant weight and sieved. For each particle size class, a subsample of 40–50 g was then processed through organic matter removal and sink-float separation (see Table 11.1). Only for Ligurian samples a separation by

Fig. 11.1 Location of the three Italian sampled sites (in green colour)

means of $CaCl_2$ solution (density 1.35 g/cm^3) was tested. In Table 11.1 sediments characteristics and different treatments used are reported too.

11.2.2.1 Organic Matter Removal

The removing of the organic substance can be performed before or after the sink-float separation. Organic matter removal can be done directly on sediments [14–16] or on the dried filter [17, 18], and are generally performed with 15% [15, 19] or 30% solution [14–16]. For Liguria beach filters, a 30% H_2O_2 solution was used. Being the 30% H_2O_2 solution too aggressive on the surface of the polymers, a 15% H_2O_2 solution was tested on Basilicata beach filters. Results have been not satisfactory because of the high amount of organic matter, therefore, a 15% H_2O_2 solution 1:1 with sediment on Calabrian beach samples was tested.

Table 11.1 Sediment characteristics and treatments used for sample analysis of the three examined Italian sites

Region	Beaches	No. of samples	Grain size distribution	Organic matter removal	Sink float separation
Liguria	Beach 1 * Beach 2	2 2	D80 4 mm D20 0.35 mm	0.5 ml 30% H_2O_2 on filter	NaCl (density 1.22 g/cm^3) (8 min for mixing and 12 h of reaction time)
Basilicata	Beach 1	4	D80 0.5 mm D20 0.1 mm	0.5 ml 15% H_2O_2 on filter >30 min of time of reaction	NaCl (density 1.22 g/cm^3) (8 min for mixing and 12 h of reaction time)
Calabria**	Beach 1* Beach 2	12 12	D80 0.88 mm D20 0.37 mm	1:1 15% H_2O_2 on whole sample, 15 min for mixing and 24 h of reaction time	NaCl (density 1.22 g/cm^3) (15 min for mixing and 24 h of reaction time)

* Lower number of visitors
** For each beach 6 samples were collected backshore and 6 foreshore

11.2.2.2 Sink-Float Separation and Filtration

Concerning the sink-float separation, the NaCl is the method more used in literature. This technique was improved for Calabria samples, mixing the sample in solution, placing it on a magnetic stirrer for 15 min, covered by an aluminium foil and letting it settle for 24 h (as suggested in [17]). From previous works, the mixing time can vary from a few seconds to 2 h [6], while the settling time can vary from 2 min to 24 h [17, 20].

The use of $CaCl_2$ solution is less used then NaCl one [21], due to the high viscosity of the solution that can obstacle the settle of organic matter. However, the high density solution can separate an higher MP amount.

After the sink-float separation, the floated material of each sample was collected from the surface of the liquid by means of a glass pipette, and filtered by a vacuum pump through a Whatman glass microfibre filter (porosity 1.2 μm, diameter 47 mm).

11.2.2.3 MP Identification Under Microscope

MPs were identified using a UV light Alonefire SV10 (365 nm, 5 W) under a Leitz ORTHOLUX II POL-MK microscope with DeltaPix Invenio 12EIII camera (Fig. 11.2, left).

To improve MP identification under UV light, the Nile Red dye was tested (Fig. 11.2, right).

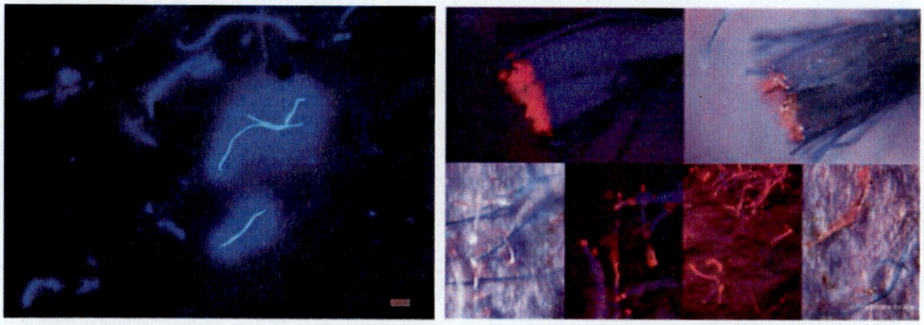

Fig. 11.2 Microplastics under microscope, using the UV light (left) and using the UV light with Nile Red dye

11.2.2.4 MP Identification with Image Analysis

Images of filters taken under UV light can be a useful instrument to identify and count MPs after a proper sample treatment and separation. Photographs of the Calabrian beach filters were taken with a smartphone (64 MP camera, resolution 9238 × 6928, focal aperture *f*/1.8) to test an automated counting software. Photos were taken with a1x zoom, brightness level of –4.0, 2 s timer and 1:1 format.

11.3 Results and Discussion

11.3.1 Organic Matter Removal

The number of MPs/kg resulting from Table 11.1 should be different due to the diverse sample treatments (see Table 11.2). The 30% H_2O_2 solution used on filter of Ligurian beaches was too aggressive on the surface of plastics particles, especially with fibres, according to previous works [22, 23].

The use of 15% H_2O_2 solution on Basilicata beach filters was not satisfactory because of the high amount of organic matter in the samples, which has not completely dissolved on the filters. Finally, the 15% H_2O_2 solution used on Calabrian beach sediments directly turned out to be the best solution.

Using $CaCl_2$ solution for MP separation, a high amount of MPs was found in the examined samples. Despite this kind of solution could be an obstacle for the settling of organic matter, it could be used to improve the methods.

Table 11.2 Microplastics abundance per kg, proportion related to grain size classes and shape in the three examined beaches

Region	Beach	MPs/kg (NaCl sink-float)	MPs/kg ($CaCl_2$ sink-float)	MP abundance in grain size	MP shape
Liguria	Beach 1* Beach 2	5340 8600	6350 4380	Both in 1–5 mm and in <0.2 mm	99% microfibers
Basilicata	Beach 1	3200	n.d.	Most abundant in finest grain size	45% microfibers, 40% microfragments, 12% microfilm, 3% microfoam
Calabria	Beach 1* Beach 2	960 910	n.d.	>1 and <0.5 mm in the same proportion	98% microfibers

* Lower number of visitors

11.3.2 Microplastic Abundance and Characterisation

In the Table 11.2, the main results obtained for the three analysed beaches are reported.

The values reported for Calabrian beaches are better specified in Table 11.3 referring to sampling performed in two different time both backshore and foreshore areas of the two beaches.

The fibers analysed by micro-FTIR were confirmed as plastic, and were mainly polyvinil alcohol and polyacrilamide.

Table 11.3 Calabria beaches MP concentration in foreshore and backshore in two different period of the year

Calabria	June 2021—MPs/kg	September 2021—MPs/kg
Beach 1—Foreshore	1000	610
Beach 1—Backshore	1090	1120
Beach 2—Foreshore	810	730
Beach 2—Backshore	780	1330

11.4 Conclusions

Approaching to MP investigation the correct use of terminology and definitions is essential to define the frame of the scientific research. To this purpose, a contrast between the scientific literature and standards has been underlined. COST project [13] and working groups are ongoing to solve this problem. In this work, the complexity of collecting and analysing real samples, the MP proper counting and recognition in the sample were discussed. For beach sediments, the grain size classification with further treatment for each class size is recommended.

Differences in MP concentration between foreshore and back shore were identified. Generally, as confirmed in literature, the backshore contained an higher number of MPs then foreshore, even if the sea movement, topologic and climatic condition can affect MP concentration.

Investigations on environmental samples confirm fibre pollution in beach sediments, not only of synthetic fibres but of artificial and natural ones. Fibre-shape and fluorescent particles represent the predominant component of MP pollution in beach sediments.

The following consideration concerning the adopted methodologies can be made:

- the grain size classification is very important to better investigate the samples;
- despite the number of MPs seems to increase in the finest class size, what is asserted by Alomar et al. in 2016 [24], which does not recognize any correlation between the abundance of microparticles and the size of the sediments, can be confirmed (Table 11.2);
- on the base of experimental results, organic matter removal with 15% H_2O_2 solution on the whole sample should be made before the sink–float separation;
- NaCl solution for sink-float separation is preferred, however, a part of MPs could be not counted. $CaCl_2$ solution could further tested to improve the separation.

References

1. De Lucia GA et al (2018) Sea water contamination in the vicinity of the Italian minor islands caused by microplastic pollution. Water 10(8):1108
2. Liu D et al (2022) Prevalence of small-sized microplastics in coastal sediments detected by multipoint confocal micro-Raman spectrum scanning. Sci Total Environ 831:154741
3. Tsang YY et al (2017) Microplastic pollution in the marine waters and sediments of Hong Kong. Mar Pollut Bull 115(1–2):20–28
4. Assas M et al (2020) Bioaccumulation and reproductive effects of fluorescent microplastics in medaka fish. Mar Pollut Bull 158:111446
5. Romeo T et al (2015) First evidence of presence of plastic debris in stomach of large pelagic fish in the Mediterranean Sea. Mar Pollut Bull 95(1):358–361
6. ISO TR 21960 Plastics—environmental aspects—state of knowledge and methodologies, 52p
7. Kannan K, Vimalkumar K (2021) A review of human exposure to microplastics and insights into microplastics as obesogens. Front Endocrinol 978

8. Hidalgo-Ruz V, Thiel M (2013) Distribution and abundance of small plastic debris on beaches in the SE Pacific (Chile): a study supported by a citizen science project. Mar Environ Res 87(88):12–18
9. Allen S et al (2019) Atmospheric transport and deposition of microplastics in a remote mountain catchment. Nat Geosci 12(5):339–344
10. Liu K et al (2019) Source and potential risk assessment of suspended atmospheric microplastics in Shanghai. Sci Total Environ 675:462–471
11. United Nations Environment Programme (2011) UNEP year book 2011: emerging issues in our global environment
12. GES Marine strategy Directive 2008/56/EC
13. Cost action plastics monitoring detection remediation recovery. https://ca-priority.eu/
14. Zhou Y, Liu X, Wang J (2019) Characterization of microplastics and the association of heavy metals with microplastics in suburban soil of central China. Sci Total Environ 694:133798
15. Zhang C et al (2019) Microplastics in offshore sediment in the yellow Sea and east China Sea, China. Environ Pollut 244:827–833
16. Mathalon A, Hill P (2014) Microplastic fibers in the intertidal ecosystem surrounding Halifax Harbor, Nova Scotia. Mar Pollut Bull 81(1):69–79
17. Balestra V, Bellopede R (2022) Microplastic pollution in show cave sediments: First evidence and detection technique. Environ Pollut 292:118261
18. Prata JC et al (2019) A new approach for routine quantification of microplastics using Nile Red and automated software (MP-VAT). Sci Total Environ 690:1277–1283
19. Panno SV et al (2019) Microplastic contamination in karst groundwater systems. Groundwater 57(2):189–196
20. Hidalgo-Ruz V et al (2012) Microplastics in the marine environment: a review of the methods used for identification and quantification. Environ Sci Technol 46(6):3060–3075
21. Schröder K, Kossel E, Lenz M (2021) Microplastic abundance in beach sediments of the Kiel Fjord, Western Baltic Sea. Environ Sci Pollut Res 28(21):26515–26528
22. Claessens M, De Meester S, Van Landuyt L, De Clerck K, Janssen CR (2011) Occurrence and distribution of microplastics in marine sediments along the Belgian coast. Mar Pollut Bull 62(10):2199–2204
23. Nuelle MT et al (2014) A new analytical approach for monitoring microplastics in marine sediments. Environ Pollut 184:161–169
24. Alomar C, Estarellas F, Deudero S (2016) Microplastics in the Mediterranean Sea: deposition in coastal shallow sediments, spatial variation and preferential grain size. Mar Environ Res 115:1–10

Chapter 12
Plastic Litter in SE Spain Beaches: Characterization and Metal Content

Joana Patrício Rodrigues, José Roberto Rivera-Hernández, Patrícia Bernárdez, Teresa Rocha-Santos, Armando Costa Duarte, and Juan Santos-Echeandía

Abstract Metal concentrations have been quantified in plastic pieces collected from four beaches located in the Mediterranean coast of Spain with different characteristics (i.e. anthropogenic pressures). Several sections of the beach were sampled (i.e. coastline, middle beach and upper dunes). After an acidic extraction, metal(oid)s levels in the samples were measured for by ICP-MS. In addition to the metal content, plastics have been classified according to their characteristics such as polymer type, color or state of degradation. These characteristics have shown to influence the adsorption capacity of metal(oid)s by plastics. Some preliminary results have shown that aged and dark colored plastics accumulate higher metal(oid)s levels probably associated to plastic surface modification. The most anthropized beaches (i.e. mining or industry) showed, in general, higher levels of metal(oid)s such as Fe, Pb and Zn. This work shows that metal content in plastics can give an idea about the pollution degree of the area where these plastic pieces have been collected. Therefore, this study is a contribution to the potential use of plastics as pollution monitors.

J. Patrício Rodrigues (✉) · T. Rocha-Santos · A. Costa Duarte
Department of Chemistry and CESAM, University of Aveiro, Campus de Santiago, 3810-193 Aveiro, Portugal
e-mail: joanarprodrigues@ua.pt

J. R. Rivera-Hernández
Instituto de Ciencias del Mar y Limnología, Universidad Nacional Autónoma de México, Unidad Académica Mazatlán, Joel Montes Camarena s/n Cerro del Vigía, 82047 Mazatlán, Sinaloa, Mexico

P. Bernárdez · J. Santos-Echeandía
Centro Oceanográfico de Vigo (IEO, CSIC), Subida a Radio Faro 50, 36390 Vigo, Spain

M. Cocca et al. (eds.), *Proceedings of the 3rd International Conference on Microplastic Pollution in the Mediterranean Sea*, Springer Water,
https://doi.org/10.1007/978-3-031-34455-8_12

12.1 Introduction

Plastic usage in land is considered a major concern for marine pollution, contributing to more than 80% of the marine litter as a result of released plastic debris [1, 2]. Literature on studies reporting the presence of metals in plastics grew in the recent years. Metals were firstly related to plastics with their inclusion as additives (i.e. catalysts, fillers, plasticizers) from their production to confer them resistance properties [3, 4]. However, in the last decades, research studies revealed that metal(oid)s can be present in polymers after being adsorbed from water [5–9]. Features of polymers may be indicative of the associated metal(oid)s, as they interfere in their uptake capacity [10]. Furthermore, properties of polymers can be naturally (degradation in the environment) or artificially (oxidized plastics) modified, conferring them a new more reactive surface [11]. Frequently, high metal(oid)s load is found in beached plastics [12, 13]. These coastal regions receive and accumulate plastic debris from macro to micro and nanoplastics both from land and sea origin. As matter of fact, these marine areas are suitable for polymer surface degradation as physical (erosion and wave action) and chemical (direct UV exposure) helped by their deposition, remaining in the sand [14]. Moreover, industrial activity is highly present in coastal areas, reuniting the ideal conditions for interaction of polymers and metals.

Color is a feature associated to specific metal addition over polymer production (e.g. copper (Cu) to blue color, and also chromium (Cr) to green color) [15]. Moreover, color of environmental plastics is frequently responsible for visual identification of aged polymers. Due to structural differences among polymers, the specific functional groups of each polymer are responsible for affinity towards certain metal and metalloid ions [10]. Furthermore, affinity can be altered or magnified through degradation from environmental exposure as well, which is driven by other physicochemical parameters. Coatings on polymer surfaces, as biofilms and mineral silts, and association with organic matter, form complexes which allow the uptake of metal(oid)s. For those reasons, it is expected to find a higher load of metal(oid)s in sampled plastics from marine environments, especially concerning more degraded pieces. The influence of the deposition area of plastics and associated metal content is also worth of investigation, as conditions differentiate.

The monitoring and control of plastics are included in environmental protection agencies, as in the European Water Framework Directive (2000/60/CE) and Marine Strategy Framework Directive (2008/56/CE), along with the reduction of metal concentrations in the environment. As far as we know, no studies were reported in literature concerning comprehensive metal(oid)s content characterization from local plastics collected in the SE Spain beaches selected for this study, highlighting the important contribution of this work.

The present study is aimed to determine the metal(oid)s content in plastics sampled from marine areas to investigate their role on estimation of local contamination degree and their potential use for metal pollution monitoring. To achieve this, metal(oid)s quantification was related to influence of: (1) anthropogenic pressure and beach zone considering different sampling areas, (2) plastic features as color, polymer and time residence on environment related to degradation. In this way, the present work will contribute for establishing important and uniform conclusions concerning the environmental risk of (micro)plastics.

12.2 Materials and Methods

12.2.1 Study Area

The selected marine areas for this study are located in the Mediterranean Coast, Murcia region, SE Spain. Portman and Calblanque are located in the Mediterranean Sea, and Islas Menores and Beal integrate the Mar Menor lagoon (Fig. 12.1). Those beaches were chosen mainly for their differences concerning anthropogenic pressure but also for other characteristics as geology, area, and related activities. Among the beaches with low human disturbance are included Calblanque, which takes part of a regional natural park, and Islas Menores, representing a remote area for mostly tourism summer season. On the contrary, Portman and Beal are recognized for being former areas for mining exploration, included in the Sierra Minera de Cartagena-La Union in the past (cessation of activity in 1990).

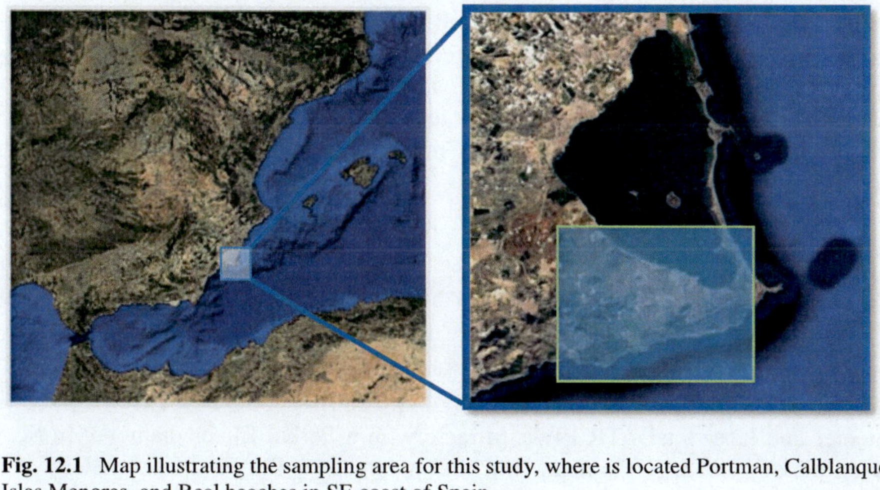

Fig. 12.1 Map illustrating the sampling area for this study, where is located Portman, Calblanque, Islas Menores, and Beal beaches in SE coast of Spain

12.2.2 Plastics Collection

Sampling took place between March–May of 2018 with manual collection of plastic pieces (less than 10 cm size) detected by naked eye, as described elsewhere [13] and following the Guideline for Monitoring Marine Litter on the Beaches from OSPAR Maritime Area [16]. The collection of plastic items was divided by 3 zones: Costal Line along 100 m (representing recent plastics deposited from the sea) to collect superficial plastics; Middle beach using density squares (50 × 50 cm area) placed randomly in the sand (×3) to collect all visible superficial sand layer plastics at maximum 1 cm depth sand; Upper zones with random collection of visible plastics over the higher beach area reenforcing a higher diversity in the plastics collected. All plastic samples were manipulated with metal free plastic material to avoid contamination. Additionally, new commercial plastics composed by different polymers were also included in this study for further metal(oid)s determination and comparison to environmental samples.

12.2.3 Screening and Classification of Plastics

Later at the laboratory, all collected samples were firstly cleaned with a brush before further manipulation, to eliminate residual organic matter and sediment particles, as much as possible, and rinsed with deionized water. All sampled plastic pieces were screened and classified to be grouped according to the following criteria: degradation appearance, beach and zone, polymer type (later confirmed by ATR-FTIR analysis) and color. Color, age, and polymer were features selected as they may influence adsorption capacity of polymers towards metal(oid)s in water. Each sample was photographed (CANON SX 410 IS with).

12.2.4 Polymer and Degradation Status Identification

Plastic pieces were cut to be analyzed by Attenuated Total Reflection Fourier-transform Infrared Spectroscopy (ATR-FTIR) (Spectrum 400 FT-IR/FT-NIR Spectrometer and horizontal ATR PIKE Miracle with a Perkin Elmer diamond/Zn Se crystal plate) to identify the main structural polymer (as PE, PP, PVC or PUR) of each sample. The methodology was described in the work of Santos-Echeandía et al. [13].

In addition, to confirm the degradation status of the observed plastics, previously characterized by naked eye, FTIR technique was applied, following Santos-Echeandía et al. [13]. Plastic samples were then classified by degradation status level (1–4) from less degraded to highly degraded.

12.2.5 Determination of Trace Metals by ICP-MS

Trace element content in the plastic samples was measured by Inductively Coupled Plasma Mass Spectroscopy (ICP-MS) using an Agilent 8900 Triple Quad with Integrated Sample Introduction System (ISIS). The following elements were monitored: magnesium (Mg), Cr, manganese (Mn), iron (Fe), cobalt (Co), nickel (Ni), Cu, zinc (Zn), arsenic (As), strontium (Sr), cadmiun (Cd), lead (Pb). Preliminary digestion of the plastic samples (200 mg) with acidic extraction was performed, as aqua regia with 3:1 HCl (20%) + HNO_3 (20%) (8 ml:2 ml) using Suprapur acids and ultrapure water. For the ICP-MS analysis, a dilution factor of 4 was applied for the samples, obtaining final sample solutions at 5% acidic mixture. The patterns for the calibration curves were prepared from ICP-MS mix stock solution and aqua regia stock solution (5%). The detection limits of the ICP-MS for trace elements were between 0.001 and 0.500 μg/g and the analytical precisions (RSD, $n = 3$) for REE and other trace elements were <10%. The analyses of the standards were in agreement with the recommended values and patterns as quality controls were analyzed when necessary to validation of the method.

12.3 Results and Discussion

12.3.1 Plastic Items

In total, 216 plastic items were considered for this study, with 201 representing beached sampled plastics, and extra 15 new commercial plastics from different uses and different polymers (Table 12.1). Portman beach had the highest number of plastics collected, particularly in the middle beach zone, followed by Calblanque. The location of both beaches in the open sea of Mediterranean coast could explain a high input of plastic debris. On the other hand, Islas Menores and Beal, in Mar Menor lagoon, presented a lower number of plastic items, which were mainly concentrated in the upper dunes. Beal and Islas Menores are mainly seasonal activity recreation beaches, which may explain the low plastic waste in low seasons, combined with regular cleaning of these two beaches all year long.

Concerning the number of plastic items by polymer, the following sequence was identified: PE > PP > PS > PUR > PVC (Fig. 12.2). This observation is in line with PE and PP being the two most globally produced polymers and, therefore, the most abundant in environment. White, blue, and colorless plastics were the predominant colors in the plastic items (14–19%), while orange, red, purple, grey, and black (2–5%) plastics were less available (Fig. 12.3).

Table 12.1 Number of plastic items collected by beach and zone and the number of new commercial plastics considered for the study

Origin	Zone	Number of items
Portman	Coastline	14
	Middle beach	42
	Upper dunes	27
Calblanque	Coastline	3
	Middle beach	35
	Upper dunes	15
Islas Menores	Coastline	3
	Middle beach	2
	Upper dunes	21
Beal	Coastline	13
	Middle beach	2
	Upper dunes	24
New commercial		15
Total		216

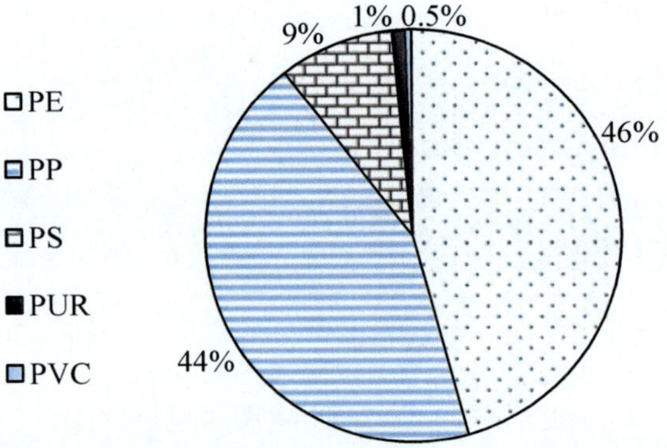

Fig. 12.2 Percentage of sampled plastics (%) according to polymer type

12.3.2 *Metal Content in Plastics*

Maximum metal(oid)s concentrations on the environmental plastic samples were sequenced as follows: Fe > Mg > Mn > Zn > Pb > Sr > As > Cu > Cr > Cd > Ni > Co (Table 12.2). Overall, Co was the metal presenting the lowest concentrations among the samples, while Fe was found with the highest values. High content of Fe on environmental polymers from beach sediment is in agreement with Maršić-Lučić et al. [17]. Pb represent the hazardous substance with the highest content estimated in the sampled plastics, despite its restriction laws in European plastic manufacturing [18]. This can be worrying, however, the high load of Pb could be justified by the presence of this element (used as additive) in legacy plastics found in environment [19]. Moreover, PVC polymer, known to have Pb in its composition [10], as observed in this study, was just represented in 0.5% of the collected samples. As matter of fact,

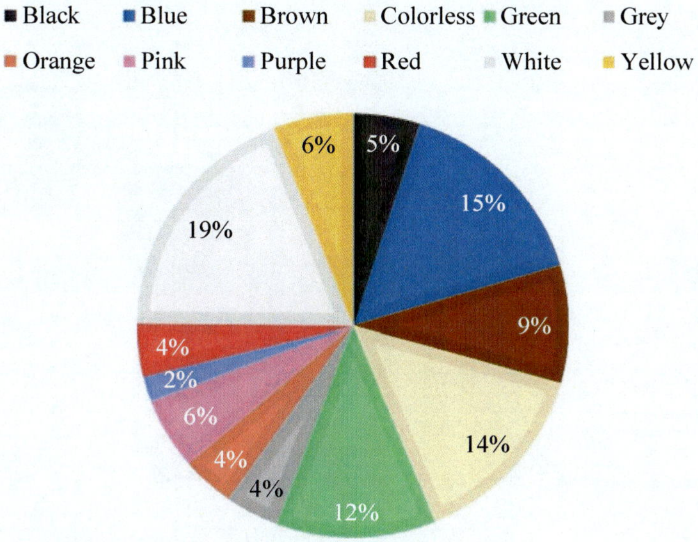

Fig. 12.3 Percentage of sampled plastic items (%) by color

Pb was quantified with low levels in the new commercial plastics used in this study. These findings may suggest that the presence of the metal Pb in the environmental plastics is due to their uptake from environment, as this metal was found in all the samples.

Table 12.2 Metal(oid)s concentrations (μg/g) range (minimum–maximum) for selected elements determined in the beached plastics collected, and maximum levels determined in the new commercial plastics considered for this study

Element	Determined min–max values in sampled plastics (μg/g)	Determined max values in new commercial plastics (μg/g)
Fe	$1.09 * 10^1$–$1.11 * 10^5$	$1.27 * 10^1$
Mg	8.89–$1.46 * 10^4$	$1.22 * 10^1$
Mn	$6.11 * 10^{-1}$–$7.29 * 10^3$	1.56
Zn	1.14–$5.67 * 10^3$	$4.08 * 10^1$
Pb	$1.24 * 10^{-1}$–$4.06 * 10^3$	$5.00 * 10^{-1}$
Sr	$2.00 * 10^{-1}$–$1.33 * 10^3$	$5.33 * 10^{-1}$
As	BDL–$6.75 * 10^2$	$1.51 * 10^{-1}$
Cu	BDL–$5.25 * 10^2$	$4.07 * 10^{-1}$
Cr	BDL–$2.78 * 10^1$	$4.06 * 10^{-1}$
Cd	BDL–$2.07 * 10^1$	$3.00 * 10^{-3}$
Ni	$2.00 * 10^{-3}$–$1.09 * 10^1$	$9.43 * 10^{-1}$
Co	BDL–$1.09 * 10^1$	BDL

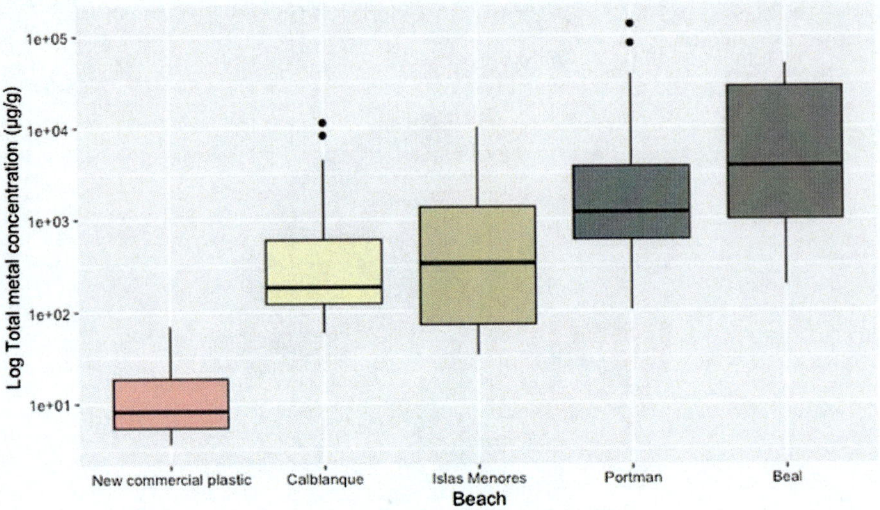

Fig. 12.4 Box plots representing the metal(oid)s content (µg/g) determined in all the plastic items by beach. Levels of metal(oid)s determined in the new commercial plastics were also included here

12.3.3 Metal Content by Plastic Classification Criteria

12.3.3.1 Beach

Substantial low levels of meta(oid)ls were determined in the new commercial microplastics (MPs), comparing to environmentally exposed plastics. Total metal(oid)s content in the selected beaches resulted as follows: Beal > Portman > Islas Menores > Calblanque (Fig. 12.4), with the highest metal(oid)s quantification identified in the most anthropogenic influenced marine areas. Overall, Beal had the highest metal(oid)s content in the sampled plastics, while Islas Menores had samples with the lowest metal(oid)s levels determined. Nevertheless, at Portman the two highest contaminated samples were found, as represented by the two high outlier values for metal(oid)s concentrations. Moreover, the outliers observed for Calblanque, which was a low contaminated beach in this study, evidence the high diversity of (micro)plastics and associated elements they carry.

12.3.3.2 Zone

For most of the metal(oid)s identified in the plastic samples, their highest content was observed in the coastline region of beaches (data not shown), representing the recently deposited plastics in the sediment from sea. This observation might be explained due to high adsorption interactions between metal(oid)s and polymers occurring in water. On the other hand, in the upper regions of the beach, degradation

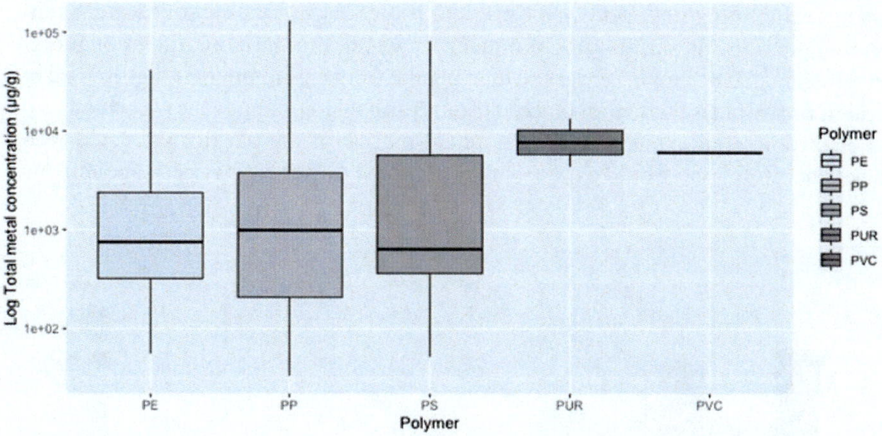

Fig. 12.5 Box plots showing the metal(oid)s load (µg/g) determined in the plastic samples by polymer type

action is more severe due to erosion (mechanical) and UV exposure (chemical), thus, some metal(oid)s can be dissociated from the plastics. Also, the upper plastics could be deposited in the sediment directly from land, avoiding contact with water where interaction with metal(oid)s essentially occurs.

12.3.3.3 Polymer

PVC and PUR based plastics contained only high levels for the determined elements in this study (Fig. 12.5), despite only 0.5 and 1%, respectively, of the plastic samples corresponded to these polymers. This observation is in line with other works [20], who reported high association of PVC polymer with metals, namely Pb, as reported before. This can be justified either with intentional addition of Pb and other elements to PVC or to its natural adsorption capacity, which could be improved due to degradation and chemical transformation, as formation of new functional groups able to bind available metal ions in water.

12.3.3.4 Color

In general, dark colored (brown and black) plastics concentrated the highest levels of metal(oid)s (Fig. 12.6), except for Cr with highest levels in red (Fig. 12.7), and also brown, plastic colors. This is in agreement with Turner and Filella [19] who attributed Cr content to brown-, red-, yellow- and orange-colored plastics. However, a few peak values in metal(oid)s concentrations were observed in individual colorless and white/yellow-colored plastics, suggesting that the abrasion of these plastics favor metal binding. Those visual characteristics may represent aged plastics which suffered from higher weathering, grinded down the original color [21], enhancing sorption capacity of polymers towards surrounding pollutants in water. Furthermore, intentional high metal load in plastics for other reason than uptake from environment is highlighted by the high metal(oid)s content in black plastics, probably coming from wasted electric and electronic equipment [22, 23], despite only 5% of the plastics collected were black. Therefore, one would expect darker plastic showing high variety of elements.

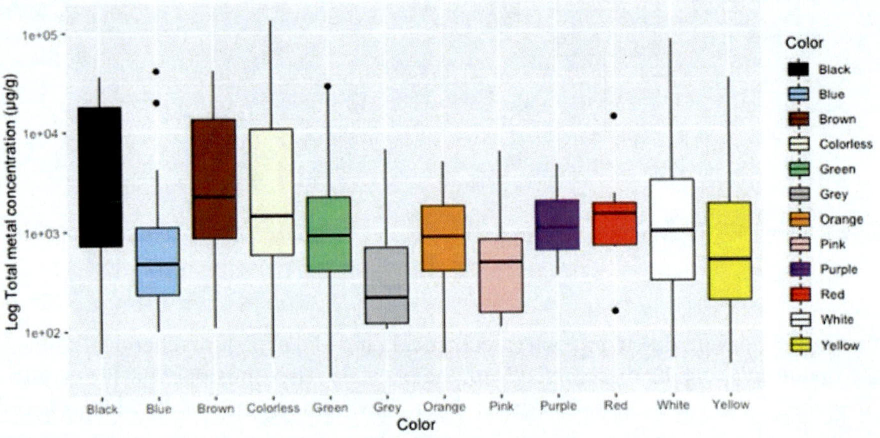

Fig. 12.6 Box plots illustrating the total metal(oid)s concentrations (µg/g) for all the plastic items by color

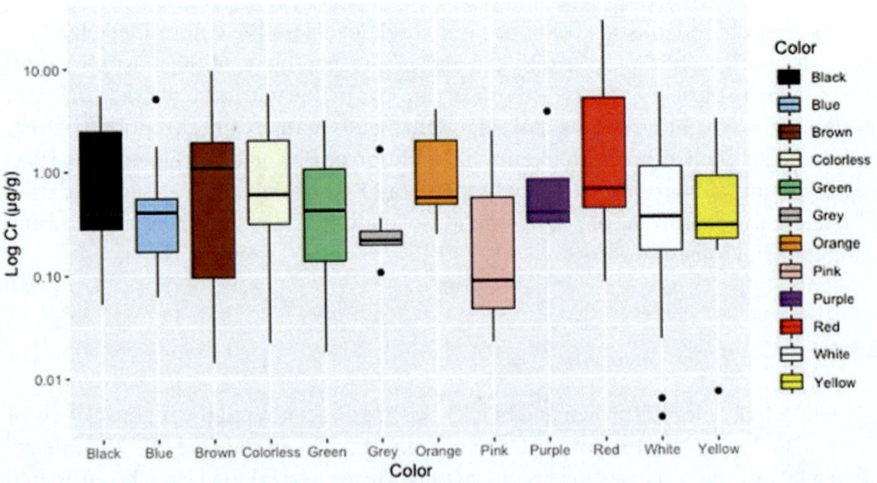

Fig. 12.7 Box plots illustrating the Cr concentrations (µg/g) considering all the plastic items by color

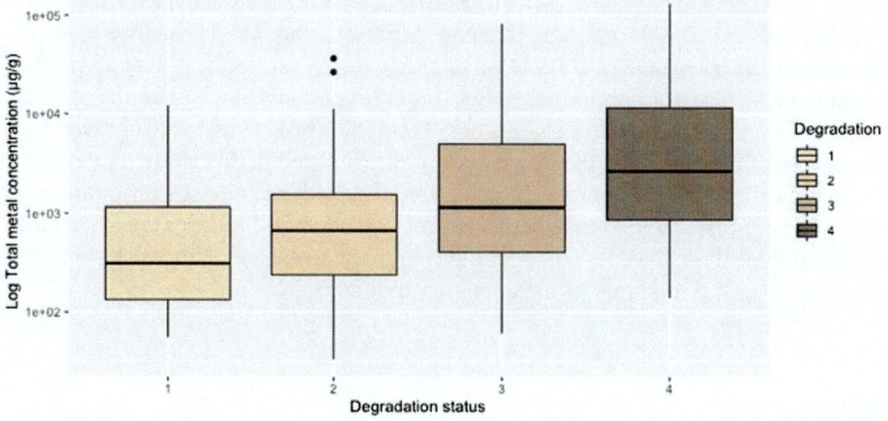

Fig. 12.8 Box plots representing the total metal(oid)s concentrations (µg/g) for all the plastic items by degradation degree

12.3.3.5 Degradation Status

From our analysis, higher metal(oid)s content was found in the most degraded (stage 3 and 4) beached plastics (Fig. 12.8). This suggest that the found elements were acquired from adsorption on environment, since weathering can remove initial additive plastics, apart from surface modification. Therefore, those observations illustrate the strong adsorption capacity of aged plastics and support degradation as a key role for uptake and storage of metal(oid)s from water, more than other characteristics or due to additives.

12.4 Conclusions

Considering all the environmental plastic samples collected from SE Spanish beaches and all the elements quantified on them, the maximum levels determined were sequenced as follows: Fe > Mg > Mn > Zn > Pb > Sr > As > Cu > Cr > Cd > Ni > Co. The major conclusions regarding our classification criteria of the items collected (by beach, zone, polymer, color and degradation status) concerning meta(oid)s concentrations is related with plastics: (1) from beaches with high anthropogenic influence (2) recently deposited in the coastal line (3) composed by PUR and PS polymers (4) darker and colorless (degraded)—indicators of aged polymers with longer time residence in environment (5) with moderated to severe degradation degree—suffering from more chemical and physical changes as oxidation, enhancing adsorption capacity of plastics towards metals.

In addition, laboratory studies are frequently performed resorting to non-realistic and higher concentrations of metal(oid)s, even concerning exposure to organisms to investigate the effects of plastic vectorization. Considering the trace levels of metals and metalloids existing in seawater, real concentrations should be primarily applied, to guarantee the relevance of a study on ecological risk assessment. Therefore, the present work is of major importance to help other researchers in the planning of their experiments under the metal(oid)s-plastics interaction investigation. Moreover, the high variety of elements and the diversity of plastics found in this study may reflect the capacity of plastics to storage (potentially adsorbed) metal(oid)s and transport them over long-distances. From our results, marine deposited plastics are suggested to reflect the local metal(oid)s content of the sampled areas, thus, this work supports plastics as potential monitors of metal environmental pollution.

References

1. GESAMP (2019) Guidelines for the monitoring and assessment of plastic litter in the ocean. In: Kershaw PJ, Turra A, Galgani F (eds) IMO/FAO/UNESCO-IOC/UNIDO/WMO/IAEA/UN/UNEP/UNDP/ISA Joint Group of Experts on the Scientific Aspects of Marine Environmental Prote, Rep. Stud. GESAMP. no 99, 130p. http://www.gesamp.org/publications/guidelines-for-the-monitoring-and-assessment-of-plastic-litter-in-the-ocean
2. Plastics Europe (2017) Plastics—the Facts 2017. https://doi.org/10.1016/j.marpolbul.2013.01.015
3. GESAMP (2015) Sources, fate and effects of microplastics in the marine environment: part 2 of a global assessment (IMO, FAO/UNESCO-IOC/UNIDO/WMO/IAEA/UN/UNEP/UNDP). In: Kershaw PJ (ed) Rep. Stud. GESAMP No. 90, 96 pp
4. Castro RO, da Silva ML, de Araújo FV (2018) Review on microplastic studies in Brazilian aquatic ecosystems. Ocean Coast Manag 165:385–400. https://doi.org/10.1016/j.ocecoaman.2018.09.013
5. Robertson DE (1968) Role of contamination in trace element analysis of sea water. Anal Chem 40:1067–1072
6. Struempler AW (1973) Adsorption characteristics of silver, lead, cadmium, zinc, and nickel on borosilicate glass, polyethylene, and polypropylene container surfaces. Anal Chem 45:2251–2254. https://doi.org/10.1021/ac60335a014
7. Giusti L, Hamilton Taylor J, Davison W, Hewitt CN (1994) Artefacts in sorption experiments with trace metals. Sci Total Environ 152:227–238. https://doi.org/10.1016/0048-9697(94)90314-X
8. Weijuan L, Youqian D, Zuyi T (2001) Americium(III) adsorption on polyethylene from very dilute aqueous solutions. J Radioanal Nucl Chem 250:497–500. https://doi.org/10.1023/A:1017953206851
9. Cobelo-Garcia A, Turner A, Millward GE, Couceiro F (2007) Behaviour of palladium(II), platinum(IV), and rhodium(III) in artificial and natural waters: influence of reactor surface and geochemistry on metal recovery. Anal Chim Acta 585:202–210. https://doi.org/10.1016/j.aca.2006.12.029
10. Rodrigues JP, Duarte AC, Santos-Echeandía J (2022) Interaction of microplastics with metal(oid)s in aquatic environments: what is done so far? J Hazard Mater Adv 6:100072. https://doi.org/10.1016/j.hazadv.2022.100072
11. Wang Q, Zhang Y, Wangjin X, Wang Y, Meng G, Chen Y (2020) The adsorption behavior of metals in aqueous solution by microplastics effected by UV radiation. J Environ Sci 87:272–280. https://doi.org/10.1016/j.jes.2019.07.006
12. Acosta-coley I, Mendez-cuadro D, Rodriguez-cavallo E, De J, Olivero-verbel J (2019)Trace elements in microplastics in Cartagena: a hotspot for plastic pollution at the Caribbean.Mar Pollut Bull 139:402–411.https://doi.org/10.1016/j.marpolbul.2018.12.016

13. Santos-Echeandía J, Rivera-Hernández JR, Rodrigues JP, Moltó V (2020)Interaction of mercury with beached plastics with special attention to zonation, degradation status and polymer type.Mar Chem 222:103788. https://doi.org/10.1016/j.marchem.2020.103788
14. Nakashima E, Isobe A, Kako S, Itai T, Takahashi S, Guo X (2016) The potential of oceanic transport and onshore leaching of additive-derived lead by marine macro-plastic debris. Mar Pollut Bull 107:333–339. https://doi.org/10.1016/j.marpolbul.2016.03.038
15. Carbery M, MacFarlane GR, O'Connor W, Afrose S, Taylor H, Palanisami T (2020) Baseline analysis of metal(loid)s on microplastics collected from the Australian shoreline using citizen science. Mar Pollut Bull 152:110914. https://doi.org/10.1016/j.marpolbul.2020.110914
16. OSPAR (2010) Guideline for monitoring marine litter on the beaches in the OSPAR Maritime Area, vol 1. OSPAR Comm., p 84
17. Maršić-Lučić J, Lušić J, Tutman P, Bojanić Varezić D, Šiljić J, Pribudić J (2018) Levels of trace metals on microplastic particles in beach sediments of the island of Vis, Adriatic Sea, Croatia. Mar Pollut Bull 137:231–236. https://doi.org/10.1016/j.marpolbul.2018.10.027
18. Hahladakis JN, Velis CA, Weber R, Iacovidou E, Purnell P (2018) An overview of chemical additives present in plastics: migration, release, fate and environmental impact during their use, disposal and recycling. J Hazard Mater 344:179–199. https://doi.org/10.1016/j.jhazmat.2017.10.014
19. Filella M, Turner A (2018) Observational study unveils the extensive presence of Hazardous elements in beached plastics from Lake Geneva. Front Environ Sci 6:1–8. https://doi.org/10.3389/fenvs.2018.00001
20. Salvaggio A, Tiralongo F, Krasakopoulou E, Marmara D, Giovos I, Crupi R, Messina G, Lombardo BM, Marzullo A, Pecoraro R, Scalisi EM, Copat C, Zuccarello P, Ferrante M, Brundo MV (2019) Biomarkers of Exposure to chemical contamination in the commercial fish species Lepidopus caudatus (Euphrasen, 1788): a particular focus on plastic additives. Front Physiol 10:1–13. https://doi.org/10.3389/fphys.2019.00905
21. Fisner M, Majer A, Taniguchi S, Bícego M, Turra A, Gorman D (2017) Colour spectrum and resin-type determine the concentration and composition of polycyclic aromatic hydrocarbons (PAHs) in plastic pellets. Mar Pollut Bull 122:323–330. https://doi.org/10.1016/j.marpolbul.2017.06.072
22. Haarman A, Gasser M (2016) Managing hazardous additives in WEEE plastic from the Indian informal sector
23. Shaw EJ, Turner A (2019) Recycled electronic plastic and marine litter. Sci Total Environ 694:133644. https://doi.org/10.1016/j.scitotenv.2019.133644

Chapter 13
Groundwater: An Endangered Invisible Resource. Microplastic Pollution in Underground Karst Systems, from Surface Watercourses to Cave Waters

Valentina Balestra, Bartolomeo Vigna, and Rossana Bellopede

13.1 Introduction

Microplastic (MP) pollution has been found in environments worldwide, including remote areas [e.g. 1, 2, 3]. MPs have a dimension less than 5 mm and are a global problem because they are easily transportable even at long distances, tend to adsorb and transport hazardous substances such as heavy metals, pesticides, persistent organic pollutants (POPs) and antibiotics [e.g. 4, 5–8], and pose a serious threat to living beings, ecosystem, and human health [e.g. 9, 10].

Despite groundwater in karst aquifers constitutes a quarter of the global drinking water sources [11], MP pollution in karst areas and underground environments is still poorly known. This fact can be related to the invisibility of subterranean karst systems and their hard access, making them difficult to study. Being open systems, karst aquifers are vulnerable to contamination by surface pollution: MP presence in groundwater can be due to human activities, wastewater, surface waters and soil pollution, atmospheric deposition and litter. The subsoil characteristics play a fundamental role in the MP transport to groundwater: MPs deposited on the ground can be transported vertically in the subsoil throughout the soil pores and rock fractures [12, 13]. Precipitations facilitate the MP infiltration into the subsoil [14], as well as

V. Balestra (✉) · B. Vigna · R. Bellopede
Politecnico Di Torino, Corso Duca Degli Abruzzi 24, Torino, Italy
e-mail: valentina.balestra@polito.it

V. Balestra · B. Vigna
Underground Karst Laboratory of Bossea Cave - S.O. Bossea C.A.I., Fraz. Fontane c/o Bossea Cave, Frabosa Soprana (CN), Italy

M. Cocca et al. (eds.), *Proceedings of the 3rd International Conference on Microplastic Pollution in the Mediterranean Sea*, Springer Water,
https://doi.org/10.1007/978-3-031-34455-8_13

transport them over long distances [15]. The hydrogeology of the study area, the flow regime of streams and their seasonal variation must be taken into account too.

Karst environments have an important role in natural ecosystems and karst aquifer are a fundamental drinking water reserve, therefore, investigations and monitoring are required for understanding the system dynamics and identify potential sources of contamination. In this work, we conducted preliminary investigations on MP pollution in different water samples from a karst area of NW Italy, considering connected surface and underground (cave) waters.

13.2 Experimental

13.2.1 Materials

Bossea cave (Frabosa Soprana, Piedmont, Italy, 836 m a.s.l.), the oldest Italian show cave, was firstly explored in the first half of the nineteenth century and opened to the public in 1874. Today, it receives about 13,000 tourist/year (2021). The cavity has a single entrance and an ascending structure, developing for about 2800 m, with a vertical drop of about 200 m. It represents the final part of the karst system developing in the Maudagna-Corsaglia watershed. The cave is fed by a system with primary (precipitations infiltrating directly into the aquifer) and secondary supply (superficial water streams losses), such as Rio Roccia bianca, the main secondary supply contributing to the Bossea karst system recharge [16]. Infiltration waters feed an important collector which cross the Bossea cave for more than 1.5 km and directly gushes out in the Corsaglia River with a set of springs [17]. A series of tests using two different dyes, previously carried out in Rio Roccia bianca, highlighted that water coming from Rio Roccia bianca arrive in the collector of Bossea cave in short time and with high concentration [16]. Different discontinuities are present in the cave, characterized by low flow circulation [17, 18].

Spot sampling were carried out in Rio Roccia bianca superficial tributary (one sample), in the Bossea cave waters (four samples), and in the Corsaglia stream, where the waters of the underground collector emerge (one sample). In the Bossea cave, two samples were collected in the speleological traits (Sifone and Polla delle anatre), as reference samples of unpolluted waters, and two in the touristic area (Uovo and Sala frane). In Sifone, in the final part of the cave, the water comes out from a siphoning tract and are collected in a lake. Polla delle anatre is a small pool of water flowing out of a fracture in the rock, entering in the inner collector [18]. Along the tourist path, in Uovo, water flows between different gours, slowing down, while, in the initial part of the cave, in Sala frane, the collector waters flow quickly before leaving the cave. Groundwater and superficial water samples were collected into 1 L glass vessels, in September 2021, under low-flow conditions, transported to the laboratory in coolers, and kept refrigerated at 6 °C until analysis.

13.2.2 Methods

MP samples were analysed according to the previously published method for cave sediments [see 19 for details], adapted to the liquid matrix. Glass filters were replaced to silver ones to determine MP typology using the infrared (IR) spectroscopy. MPs were detected by exploiting their fluorescence under ultraviolet (UV) light, given by Fluorescent Whitening Agents (FWAs), commonly used additives in plastic production [20, 21]. All researchers used nitrile gloves and cotton coats. For every step, all surfaces and materials were cleaned with ethanol and distilled water to avoid external contaminations. Plastic equipment was replaced with glass or metal one.

13.2.2.1 Preparation of Samples

Water samples were filtered through a 0.8 μm pore size silver filter (GVS Life Sciences, Membrane Disk 47 mm). Filters, positioned on glass petri dishes and covered with aluminium foil, were dried in an oven at 40 °C for 2 h. The organic matter removal was carried out on dried filters through the application of 2 ml of 15% hydrogen peroxide solution, left to react for 30 min at room temperature and, finally, dried again for 1 h at 50 °C.

13.2.2.2 Analytical Techniques

Filters were observed with and without an UV flashlight (Alonefire SV10 365 nm UV flashlight 5W) under a Leitz ORTHOLUX II POL-MK microscope equipped with a DeltaPix Invenio 12EIII 12 Mpx Camera, with a 2.5x, 4x, 10 × or higher magnifications. Images of natural, artificial and synthetic fibres under microscope [e.g. 22, 23, 24] were used to compare particles observed under microscope, especially transparent fibres. The UV flashlight was positioned with an inclination of 45 °C. Visual identification was used to count MPs, according to strict selection criteria described in previous research [25–27]. A cut off of the particles smaller than 0.1 mm was made as suggested in European Commission [28]. Particles that could not be identified as MPs were not take into consideration. The counted MPs were described using the Standardised size and colour sorting system (SCS) [25].

Finally, a 10% of MPs of each filter were analysed using a micro-Fourier Transform Infrared Spectroscope (micro-FTIR) Shimadzu AIM-9000 microscope equipped with a Shimadzu IRTracer-100 spectrophotometer. Spectra were compared with the Shimadzu standard library.

13.3 Results and Discussion

13.3.1 Results

MPs were found in all water samples: 54 items/L in Sifone, 16 items/L in Sala frane, 12 items/L in Polla delle anatre, 30 items/L in Uovo, 23 items/L in Rio Roccia bianca and 29 items/L in Corsaglia River. Fibres represented the majority of the MPs present in the karst system (95.1%), followed by fragments (4.3%) and beads (0.6%). Mini-MPs accounted for 82.9% of the total MPs found in water karst system. The highest MP abundance was fluorescent under UV light (84.8%). Of the fluorescent particles, 46% were transparent, 16.5% red, 10.8% blue, 7.2% beige, and 26.7% other colours. Non-fluorescent MPs were mainly black (68%) or blue (20%). Polyethylene (51.4%) and polyvinyl alcohol (31.4%) are the main types of MPs found in the karst system waters, followed by polyester (8.9%), ethylene vinyl alcohol (2.9%), polyvinyl chloride (2.9%) and acrylic adhesive (2.9%).

13.3.2 Discussion

At the moment, a standardized method for assessing MPs do not exist. Using different methodology makes it difficult to compare the results. However, some considerations on MP amount in karst environment can be equally done. The only work detecting MP pollution in springs and wells in a karst water area has documented a maximum value of 15.2 MPs/L, under low-flow conditions [11], a fairly low value compared to those found in the Bossea karst system. Instead, concerning MP characteristics, the data about MPs in cave waters are similar to those described in Balestra and Bellopede [19] for Bossea cave sediments. Five sediment samples were collected along different part of the tourist path and one in the not-tourist area of the Bossea cave. Comparisons between the mean values obtained from sediment and water samples analysis inside Bossea cave are reported in Figs. 13.1, 13.2. The major part of MPs in Bossea cave waters are fluorescent, fibre-shape, and have a dimension between 0.99 and 0.1 mm, as in cave sediments (Fig. 13.1). The most of the fluorescent particles in Bossea cave waters are transparent, as in cave sediments, and the other main colours are similar too, with except of red particles, greater in water (Fig. 13.2). The non-fluorescent MPs were mainly black and blue, confirming the data found in Bossea cave sediments (Fig. 13.2). However, percentages of these data are different, as the colour of the less abundant MPs (Fig. 13.2). It could be related to different kind of biodeterioration, or to the diverse density of certainly MPs, more easily transportable in liquid medium. MPs in groundwater of Bossea karst system could be related to outside pollution, linked to the air transport of particles, soil contamination, winter activities related to the ski resorts and pollution from the near countries and could be transported especially by the main secondary supplies such as Rio Roccia bianca. The waters in the touristic part of the cave could be enriched with pollutants too, brought inside by

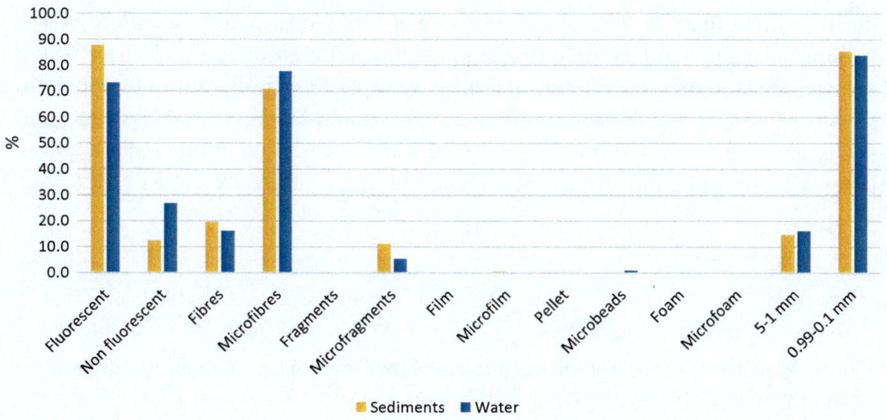

Fig. 13.1 Comparison of size, shape and fluorescence between microplastics found in Bossea cave waters and sediments

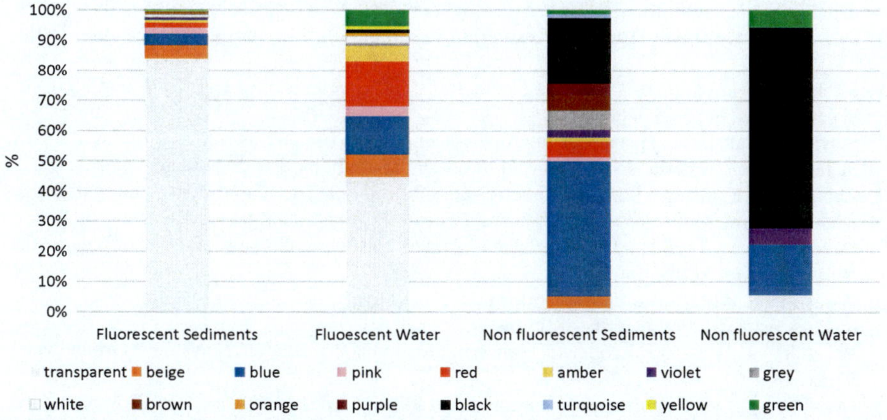

Fig. 13.2 Comparison of colours between fluorescent and non-fluorescent microplastics found in Bossea cave waters and sediments

tourists [19], which could be transported in different part of the cave and outside, in the Corsaglia River.

13.4 Conclusions

This study documented the presence of microplastics in karst systems, from surface watercourses to water in cave. Microplastics with a dimension less than 1 mm and fibre shape dominate the samples, and polyethylene is the main types of microplastics found in the karst system waters. Fractured aquifers may be vulnerable to microplastic pollution, contaminating waters and nearby environments, therefore, MP detection is essential to understand the pollution amount and define strategies for conservation. Moreover, karst aquifers are open systems, susceptible to pollutant contamination from surface, therefore, the above-ground areas should be monitored and protected too. Greater endeavours should be done to karst areas protection, implementing pollutants monitoring and sustainably managing important resources as water.

References

1. Zhang Y et al (2021) Microplastics in glaciers of the Tibetan Plateau: Evidence for the long-range transport of microplastics. Sci Total Environ 758:143634
2. Ambrosini R et al (2019) First evidence of microplastic contamination in the supraglacial debris of an alpine glacier. Environ Pollut 253:297–301
3. Cincinelli A et al (2017) Microplastic in the surface waters of the Ross Sea (Antarctica): occurrence, distribution and characterization by FTIR. Chemosphere 175:391–400
4. Li X et al (2019) Enhancement in adsorption potential of microplastics in sewage sludge for metal pollutants after the wastewater treatment process. Water Res 157:228–237
5. Li J, Zhang K, Zhang H (2018) Adsorption of antibiotics on microplastics. Environ Pollut 237:460–467
6. Rochman CM et al (2013) Ingested plastic transfers hazardous chemicals to fish and induces hepatic stress. Sci Rep 3(1):1–7
7. Selvam S et al (2021) Hazardous microplastic characteristics and its role as a vector of heavy metal in groundwater and surface water of coastal south India. J Hazard Mater 402:123786
8. Zhou Y, Liu X, Wang J (2019) Characterization of microplastics and the association of heavy metals with microplastics in suburban soil of central China. Sci Total Environ 694:133798
9. Devereux R et al (2021) The abundance of microplastics in cnidaria and ctenophora in the North Sea. Mar Pollut Bull 173:112992
10. Assas M et al (2020) Bioaccumulation and reproductive effects of fluorescent microplastics in medaka fish. Mar Pollut Bull 158:111446
11. Panno SV et al (2019) Microplastic contamination in karst groundwater systems. Groundwater 57(2):189–196
12. Chia RW, et al. (2021) Microplastic pollution in soil and groundwater: a review. Environmental Chemistry Letters, 1–14
13. Wanner P (2021) Plastic in agricultural soils–a global risk for groundwater systems and drinking water supplies?–a review. Chemosphere 264:128453
14. Zhou Y et al (2021) Microplastic contamination is ubiquitous in riparian soils and strongly related to elevation, precipitation and population density. J Hazard Mater 411:125178
15. Allen S et al (2019) Atmospheric transport and deposition of microplastics in a remote mountain catchment. Nat Geosci 12(5):339–344
16. Vigna, B (2020) Assetto geologico ed idrogeologico del Sistema carsico di Bossea (SW Piemonte, Italy), in Atti del Convegno Nazionale "L'uomo domanda, la grotta risponde". Cinquantesimo anniversario del Laboratorio Carsologico Sotterraneo di Bossea, Frabosa Soprana (CN), Italy. p 283–300
17. Vigna B, (2021) Vadose Zone Hydrogeology In The Bossea Cave System (Southern Piedmont, Northern Italy), in Proceedings of the 17th International Congress of Speleology, Sidney. p 222–225
18. Balestra V, Fiorucci A, Vigna B (2022) Study of the trends of chemical-physical parameters in different karst aquifers: some examples from italian alps. Water 14(3):441
19. Balestra V, Bellopede R (2022) Microplastic pollution in show cave sediments: First evidence and detection technique. Environ Pollut 292:118261

20. Ehlers SM, Maxein J, Koop JH (2020) Low-cost microplastic visualization in feeding experiments using an ultraviolet light-emitting flashlight. Ecol Res 35(1):265–273
21. Klein M, Fischer EK (2019) Microplastic abundance in atmospheric deposition within the Metropolitan area of Hamburg. Germany. Sci. Total Environ. 685:96–103
22. Houck MM (2009) Identification of textile fibers
23. Khan A, et al. (2017) A review paper on Textile Fiber Identification. IOSR Journal of Polymer and Textile Engineering (IOSR-JPTE) 4, 14–20
24. Zhang, X (2014) Fundamentals of fiber science
25. Crawford CB, Quinn B (2016) Microplastic pollutants, Amsterdam
26. Noren F (2007) Small plastic particles in coastal swedish waters, in n-research report, commissioned by KIMO, Sweden
27. Hidalgo-Ruz V et al (2012) Microplastics in the marine environment: a review of the methods used for identification and quantification. Environ Sci Technol 46(6):3060–3075
28. European Commission: Guidance on monitoring of marine litter in European seas. A guidance document within the common implementation strategy for the Marine Strategy Framework Directive. Ispra: European Commission, Joint Research Centre, MSFD Technical Subgroup on Marine Litter. p 126 (2013)

Chapter 14
Evaluation of the Role of Plastic Aging on the Sorption of Pesticides and Pharmaceutical Substances on Microplastics

Mariana N. Miranda, Adrián M. T. Silva, and M. Fernando R. Pereira

Abstract Microplastics' sorption capability has been studied over the last decade to better understand the potential impacts that they have on the ecosystems and human health. However, the role of plastic aging, in which the polymer's physical and chemical properties are changed due to exposure to environmental agents, such as weathering, has been overlooked. Consequently, there is a need to explore the role of the aging processes in the interaction of microplastics with other pollutants, and the potential risk of an enhanced capacity of microplastics carrying co-occurring pollutants between environmental compartments and ecosystems. Thus, the main goal of this study was to evaluate the changes in the sorption capacity after aging microplastics of LDPE – low-density polyethylene, PET – poly(ethylene terephthalate), and uPVC – unplasticized poly(vinyl chloride). Through sorption experiments, virgin and aged (by ozone exposure or 3 months of rooftop weathering) microplastic particles of the three polymers were exposed to ten organic contaminants (pesticides or pharmaceutical substances) at trace concentrations, including both priority substances and contaminants of emerging concern. The results show increased sorption of these contaminants on the aged microplastic particles, which is dependent on the affinity between each polymer and organic contaminant, and the effectiveness of the aging treatment. A better understanding of the interaction between these different pollutants was reached due to the analysis of the sorption experiments results when accounting for modifications of the chemical structure, surface morphology and/or crystallinity of the microplastic particles identified after the aging processes. This study supports the hypothesis that microplastics can be transport vectors for other pollutants and demonstrates the major role of the aging degree in the sorption process.

M. N. Miranda (✉) · A. M. T. Silva · M. F. R. Pereira
LSRE-LCM – Laboratory of Separation and Reaction Engineering – Laboratory of Catalysis and Materials, Faculty of Engineering, University of Porto, Rua Dr. Roberto Frias, 4200-465 Porto, Portugal
e-mail: mmiranda@fe.up.pt

ALiCE – Associate Laboratory in Chemical Engineering, Faculty of Engineering, University of Porto, Rua Dr. Roberto Frias, 4200-465 Porto, Portugal

M. Cocca et al. (eds.), *Proceedings of the 3rd International Conference on Microplastic Pollution in the Mediterranean Sea*, Springer Water,
https://doi.org/10.1007/978-3-031-34455-8_14

Thus, it provides further evidence of the pressing need to study microplastics in more realistic conditions, in which they undergo aging and interact with other pollutants.

14.1 Introduction

Microplastics are considered as contaminants of emerging concern [1] since large quantities started to be frequently and in a widespread manner found in several environmental compartments and ecosystems [2]. In order to better understand the behavior and impacts of microplastics, their sorption capacities started to be studied, but the role of plastic aging on the sorption processes has been overlooked [3]. Polymer degradation, by abiotic and/or biotic pathways, is usually a slow process under natural environmental conditions, which creates physical and chemical modifications, such as crazing, cracking, the introduction of oxygen-containing surface groups, and discoloration [4]. These modifications are expected to change the behavior of microplastics over time, influencing their interaction with co-occurring contaminants and biota [5, 6]. Consequently, there is a need to explore the role of the aging processes in the interaction of microplastics with other pollutants, and the subsequent potential risk of an enhanced sorption capacity and a trojan horse vector effect [7].

Although several pesticides and pharmaceutical substances are already considered priority substances under the European Union (EU) environmental legislation, new substances continue to be developed and pose new threats to the environment. These substances, yet to have Environmental Quality Standards [8], are classified as contaminants of emerging concern. Therefore, research is necessary on the interaction of these substances with microplastics to identify affinities between the different polymers and substances and the potential threats of this combo in the environment [9].

Thus, the goals of this study were to (i) assess the modifications on the chemical structure, surface morphology and crystallinity of microplastic particles of 3 polymers after different aging treatments; (ii) evaluate the changes in the sorption capacity after aging and link those with the modifications detected, and (iii) identify affinities between 10 organic contaminants (4 pesticides and 6 pharmaceutical substances) with microplastic particles of the selected polymers.

14.2 Experimental

14.2.1 Materials

Microplastic particles (MPPs) of LDPE – low-density polyethylene (average particle diameter of 509 ± 221 μm), PET – poly(ethylene terephthalate) (161 ± 79 μm) and uPVC – unplasticized poly(vinyl chloride) (159 ± 43 μm) were purchased from Goodfellow (UK).

Ten organic contaminants were selected for the study, including both priority substances and contaminants of emerging concern, at trace concentrations: (i) pesticides – alachlor, clofibric acid, diuron, pentachlorophenol; and (ii) pharmaceutical substances – citalopram, diclofenac, florfenicol, tramadol, trimethoprim, and venlafaxine. These substances were purchased from Sigma-Aldrich (Germany) or Supelco (Germany) in powder form.

14.2.2 Methods

The MPPs were first aged by ozone exposure or rooftop weathering, being characterized before and after aging. Then, sorption batch experiments were conducted with both virgin and aged samples of MPPs. Finally, the solutions were filtered to remove the MPPs from suspension and the pesticides and pharmaceutical substances in solution were quantified by liquid chromatography.

14.2.2.1 Aging Treatments

For the ozone gas MPP aging, the virgin MPPs of each polymer received a 150 $Ncm^3\ min^{-1}$ flow rate of 50 $gO_3\ Nm^{-3}$ for 1 h [10]. For the rooftop (RT) weathering experiment, the virgin MPPs of each polymer were exposed to real conditions of weathering regarding air humidity and temperature, sunlight, and precipitation (under a temperate climate, during the period of the year with the maximum hours of sunlight per day and higher UV index, in the city of Porto, Portugal). Two types of weathering samples were obtained after 3 months (13 weeks): RT1 – the MPPs underwent aging due to natural solar exposure and temperature variations; RT2 – same as RT1 but filtered rainwater collected was additionally added to these samples of MPPs, following the precipitation pattern [10].

14.2.2.2 Sorption Experiments

Two series of sorption batch experiments were performed to study the MPP sorption capacities and affinity to the organic contaminants [11]: (i) sorption kinetics of a mix of the 10 organic contaminants on virgin MPPs; (ii) sorption with the same mix of the 10 organic contaminants but using aged MPPs to study this effect.

For this, 250 mg of MPPs (LDPE, PET, or uPVC) and 25 mL of the solution with the 10 organic contaminants (pH ca. 5.8, ca. 100 μg L^{-1} of each substance) were added to 50 mL bottles. The bottles were shaken at 150 strokes per minute and kept at 25 °C for the duration of the experiment. All samples with MPPs were run in triplicates and had control bottles without the MPPs. After that, the samples underwent vacuum filtration (pore size of 1.2 μm) to remove the MPPs from suspension. The filtered samples were stored in HPLC vials at −18 °C until analyzed.

For the kinetics experiments, the following time intervals were used: 2, 4, 8, 16, 24 and 30 h. For the other type of experiments, the samples were shaken for 24 h.

14.2.2.3 Analytical Techniques

The MPPs were characterized using a particle size analyzer, Attenuated Total Reflection – Fourier transform infrared (ATR-FTIR) spectroscopy, Scanning Electron Microscopy (SEM), and X-ray diffraction (XRD) analysis [10].

The organic substances were quantified using an ultra-high-performance liquid chromatography-tandem mass spectrometry (UHPLC-MS/MS) analytical method [11].

14.2.2.4 Data Analysis

Carbonyl Index (*CI*) was estimated by Eq. 14.1 for LDPE and uPVC [10]. The ATR-FTIR spectra were used to obtain the transmittance, which was converted to absorbance (*A*), of the carbonyl group and methylene group peaks maximum height. *CI* was not calculated for PET, since this polymer has a characteristic infrared absorption peak at 1714 cm^{-1} due to C = O stretching of the carboxylic acid group, disabling the possible detection of the growth of the carbonyl group peak height as aging modifies the polymer. The average and standard deviation were calculated for all the results.

$$CI = \frac{A_{\text{carbonyl group(1712–1722)}}}{A_{\text{methylene group(2848 for LDPE or 2844 for uPVC)}}} \tag{14.1}$$

The XRD diffractograms were analyzed using the HighScore 4.8 software, in this way determining the crystalline and amorphous areas and, consequently, the degree of crystallinity [10].

The sorption capacity q (μg g^{-1}) was calculated by Eq. 14.2, where C_{control} is the concentration of each organic substance (μg L^{-1}) in the control bottle liquid phase, C_{sample} is the concentration of each organic substance (μg L^{-1}) in a sample bottle liquid phase, V is the volume (L) of the solution added to each bottle, and m is the mass (g) of MPPs added to the bottle.

$$q = \frac{(C_{\text{control}} - C_{\text{sample}})}{m} V \tag{14.2}$$

14.3 Results and Discussion

Of the 3 polymers selected, LDPE was the most impacted by the aging treatments studied Fig. 1a. Both the ozone treatment and rooftop weathering lead to modifications in the chemical structure of the polymer. The most substantial modification was the formation of carbonyl species (C = O), enabling to quantify the degree of aging with the estimation of the *CI* increase from 0.004 ± 0.001 (virgin) to 0.100 ± 0.002 (ozone), 0.243 ± 0.011 (RT1) and 0.246 ± 0.006 (RT2) [10]. While being more difficult to assess the aging degree of PET using ATR-FTIR data, it was possible to detect a broader 1714 cm^{-1} peak base for the ozone samples (Fig. 1b). The weathering treatments were the most effective in introducing changes in the chemical structure of uPVC, as suggested by the highest increase in carbonyl species Fig. 1c. The uPVC *CI* increased from 0.73 ± 0.15 (virgin) to 0.85 ± 0.06 (ozone), 4.82 ± 0.27 (RT1) and 3.37 ± 0.12 (RT2) [10]. Another noteworthy modification identified is the formation of hydroxyl species (O–H) for the weathering samples of the 3 polymers, although much more subtle. Overall, LDPE was shown to be more susceptible, than PET and uPVC, to crystallinity changes under the aging treatments conducted and for the exposure times tested, particularly after weathering.

The SEM analysis of the surface morphology of LDPE, PET and uPVC revealed that most of the aging treatments lead to no major morphological changes. The most relevant modifications detected occurred for LDPE after weathering, as shown by the observation of randomly distributed cracks on the surface of LDPE weathered samples, that were more pronounced for the RT2 samples, which were in contact with rainwater [10].

The sorption kinetics of the mixture of the 10 organic contaminants were studied for each polymer virgin MPPs, allowing to define an adequate contact time for the rest of the sorption experiments. These first experiments show that only alachlor (for LDPE) and pentachlorophenol (for the 3 polymers but mainly for LDPE) were discernibly sorbed on virgin MPPs. An apparent equilibrium was reached between 16 and 24 h. Thus, 24 h was considered enough to reach equilibrium, under the conditions tested.

Following this, the effect of aging on the sorption capacities of the polymers was studied (Fig. 14.2). The results revealed higher sorption of several organic substances on aged MPPs, when compared to virgin MPPs, i.e. the sorption capacity increased from one or two sorbed substances (maximum 3 $\mu g\ g^{-1}$ per sorbate) up to nine after aging (maximum 10 $\mu g\ g^{-1}$ per sorbate). As predicted, considering the characterization of the aged samples, LDPE sorption capacities were the most impacted by aging. The weathering exposure led to the highest increases in the sorption capacities

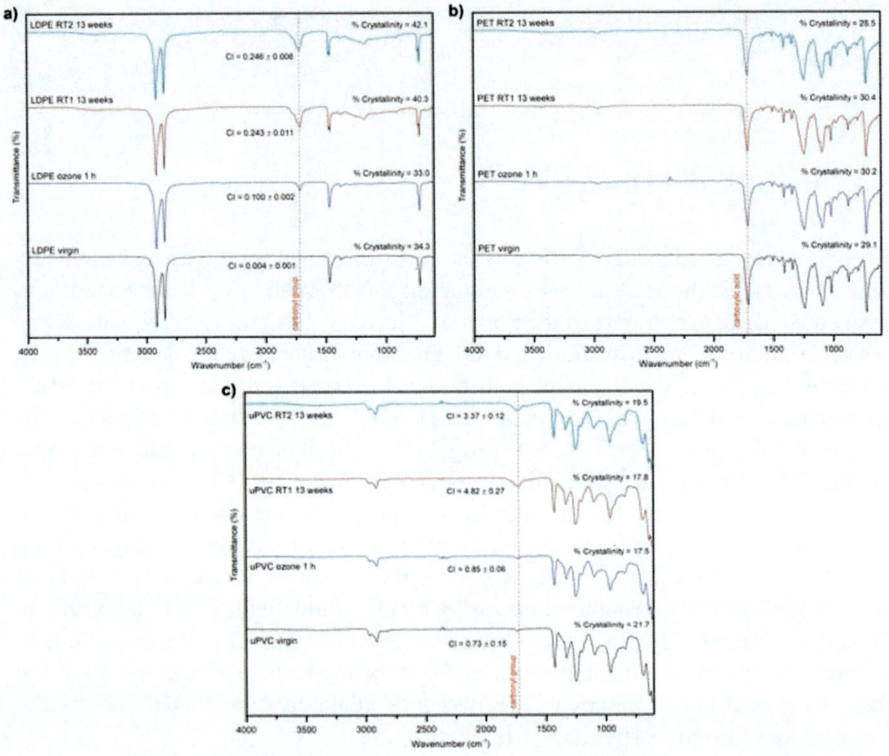

Fig. 14.1 Microplastic particles (MPPs) ATR-FTIR spectra, Carbonyl Index (*CI*) and degree of crystallinity: **a** LDPE, **b** PET, and **c** uPVC

of LDPE and uPVC, but ozone exposure proved to be more effective for PET due to pentachlorophenol sorption on the ozone-aged MPPs of this polymer.

The increase in the amount of the oxygen-containing functional groups through aging is known for affecting the surface hydrophobic properties of the plastics and allowing additional interactions [12]. Thus, the previously identified modifications can be linked to the increment of the sorption capacity for certain substances. For instance, the overall higher sorption capacity observed for uPVC RT1 samples over RT2 samples matches the higher *CI* obtained for RT1 over RT2. Likewise, the similar *CI* obtained for RT1 and RT2 samples of LDPE matches the similar sorption capacities between these two samples.

Despite the fact that RT1 and RT2 samples of LDPE show similar sorption capacities for most substances and similar modifications in their chemical structure, they are different regarding their surface morphology, i.e. RT2 is more aged in that regard. This suggests that the affinity and consequent sorption of most of the studied substances on LDPE MPPs were mainly governed by the surface chemical structure.

Even after aging, pentachlorophenol remains the most sorbed substance for all LDPE samples and ozone-aged PET. Another noteworthy affinity encountered was

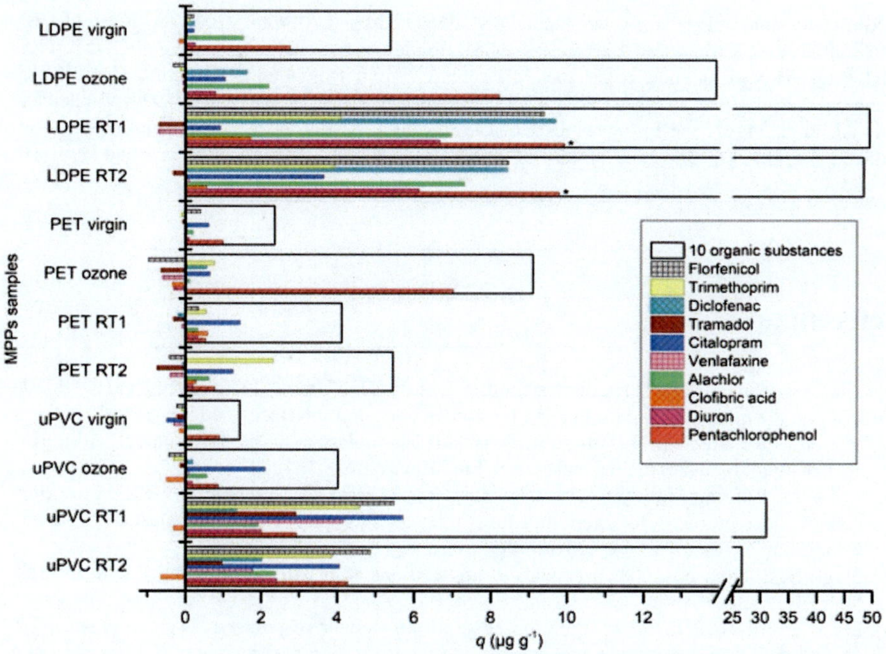

Fig. 14.2 Microplastic particles (MPPs) sorption capacity for organic contaminants. * Pentachlorophenol concentration in solution after 24 h was below the method detection limit for LDPE RT1 and RT2 samples, and therefore q was calculated admitting a null concentration for those samples

the trimethoprim sorption on the PET RT2 samples, which might be related to the formation of hydroxyl species during aging. No affinity was found, even after aging, between tramadol and venlafaxine for LDPE and PET, and between clofibric acid and uPVC.

14.4 Conclusions

This study demonstrates the major role of the aging degree of MPPs in the sorption process, with impact on the interaction between most of the studied organic contaminants and the MPPs of LDPE, PET and uPVC. The differences between the sorption capacities of each polymer before and after aging can be linked to the chemical and physical modifications identified, providing essential information on the behavior of microplastics in the environment, particularly when predicting their affinity with a diversity of co-occurring contaminants. For LDPE and uPVC, the *CI* can be a valuable indicator in a preliminary estimation of the global sorption capacities of aged MPPs, although it is not valid for every single substance.

Acknowledgments This work was financially supported by: LA/P/0045/2020 (ALiCE), UIDB/50020/2020 and UIDP/50020/2020 (LSRE-LCM), funded by national funds through FCT/MCTES (PIDDAC). M.N.M. acknowledges the PhD research grant from Fundação para a Ciência e a Tecnologia (FCT) (Ref. PD/BD/137730/2018, extended by COVID/BD/152633/2022), funded by national funds and by the European Union (EU) through the European Social Fund (ESF). The authors are grateful for the contributions of Ana R. Lado Ribeiro, Maria J. Sampaio, and Pedro B. Tavares.

References

1. The European Parliament and the Council of the European Union (2020) Directive (EU) 2020/2184 on the quality of water intended for human consumption (recast) 435:1–62
2. Miranda MN, Silva AMT, Pereira MFR (2020) Microplastics in the environment: A DPSIR analysis with focus on the responses. Sci Total Environ 718:134968
3. Alimi OS, Claveau-Mallet D, Kurusu RS, Lapointe M, Bayen S, Tufenkji N (2022) Weathering pathways and protocols for environmentally relevant microplastics and nanoplastics: What are we missing? J Hazard Mater 423:126955
4. Singh B, Sharma N (2008) Mechanistic implications of plastic degradation. Polym Degrad Stab 93(3):561–584
5. Liu P et al (2020) Effect of aging on adsorption behavior of polystyrene microplastics for pharmaceuticals: Adsorption mechanism and role of aging intermediates. J Hazard Mater 384:121193
6. Vroom RJE, Koelmans AA, Besseling E, Halsband C (2017) Aging of microplastics promotes their ingestion by marine zooplankton. Environ Pollut 231:987–996
7. UNEP (2016) Marine plastic debris and microplastics: Global lessons and research to inspire action and guide policy change. Nairobi. Available: https://wedocs.unep.org/handle/20.500.11822/7720
8. The European Parliament and the Council of the European Union (2013) Directive 2013/39/EU of the European Parliament and of the Council of 12 August 2013 amending Directives 2000/60/EC and 2008/105/EC as regards priority substances in the field of water policy 226:1–17
9. Vieira Y, Lima EC, Foletto EL, Dotto GL (2021) Microplastics physicochemical properties, specific adsorption modeling and their interaction with pharmaceuticals and other emerging contaminants. Sci Total Environ 753:141981
10. Miranda MN, Sampaio MJ, Tavares PB, Silva AMT, Pereira MFR (2021) Aging assessment of microplastics (LDPE, PET and uPVC) under urban environment stressors. Sci Total Environ 796:148914
11. Miranda MN, Lado Ribeiro AR, Silva AMT, Pereira MFR (2022) Can aged microplastics be transport vectors for organic micropollutants? – Sorption and phytotoxicity tests. Sci Total Environ 850:158073
12. Liu G, Zhu Z, Yang Y, Sun Y, Yu F, Ma J (2019) Sorption behavior and mechanism of hydrophilic organic chemicals to virgin and aged microplastics in freshwater and seawater. Environ Pollut 246:26–33

Chapter 15
Microplastics: What Can We Learn from Clastic Sediments?

Kryss Waldschläger, Muriel Z. M. Brückner, Bethanie Carney Almroth, Christopher R. Hackney, Tanveer M. Adyel, Olubukola S. Alimi, Sara L. Belontz, Win Cowger, Darragh Doyle, Andrew Gray, Ian Kane, Merel Kooi, Matthias Kramer, Simone Lechthaler, Laura Michie, Tor Nordam, Florian Pohl, Catherine Russell, Amalie Thit, Wajid Umar, Daniel Valero, Arianna Varrani, Anish K. Warrier, Lucy C. Woodall, and Nan Wu

15.1 Introduction

Microplastics research is a comparatively young discipline that only gained momentum at the beginning of the twenty-first century. Research on clastic sediment, on the other hand, has been conducted for almost a hundred years [1–3] and provides valuable insights to guide future microplastic studies. Comparing publication numbers in the Web of Science also highlights the head start that sediment

K. Waldschläger (✉)
Hydrology and Quantitative Water Management Group, Wageningen University & Research, Wageningen, Netherlands
e-mail: kryss.waldschlager@wur.nl

M. Z. M. Brückner
College of Life and Environmental Sciences, University of Exeter, Exeter, UK

B. C. Almroth
Department of Biology and Environmental Sciences, University of Gothenburg, Gothenburg, Sweden

C. R. Hackney
School of Geography, Politics and Sociology, Newcastle University, Newcastle-Upon-Tyne, UK

T. M. Adyel
STEM, University of South Australia, Mawson Lakes, Australia

O. S. Alimi
Department of Chemical Engineering, McGill University, Montreal, Canada

S. L. Belontz
Department of Earth Sciences, University of Western Ontario, ON, Canada

M. Cocca et al. (eds.), *Proceedings of the 3rd International Conference on Microplastic Pollution in the Mediterranean Sea*, Springer Water,
https://doi.org/10.1007/978-3-031-34455-8_16

research (*sediment* transport*: 62,617 publication) has over microplastic research (*microplastic* transport*: 1352 publications). Based on the hypothesis that publications equal knowledge, we have analyzed the comparability of microplastics with natural sediments regarding their particle properties, transport processes, sampling techniques and ecotoxicology in an interdisciplinary review paper [4]. The paper

W. Cowger
Department of Environmental Sciences, University of California, USA AND Moore Institute for Plastic Pollution Research, Riverside, USA

D. Doyle
Department of Biology and Environmental Sciences, University of Gothenburg, Gothenburg, Sweden

A. Gray
Department of Environmental Sciences, University of California, Riverside, USA

I. Kane
Department of Earth and Environmental Sciences, University of Manchester, Manchester, UK

M. Kooi
Aquatic Ecology and Water Quality Management Group, Wageningen University & Research, Wageningen, the Netherlands

M. Kramer
School of Engineering and Information Technology (SEIT), UNSW Canberra, Canberra ACT 2610, Australia

S. Lechthaler
Institute of Environmental Engineering, RWTH Aachen University, Aachen, Germany

L. Michie
Institute of Marine Sciences, University of Portsmouth, Ferry Road, Portsmouth PO4 9LY, UK

T. Nordam
Norway and Department of Physics, Norwegian University of Science and Technology, SINTEF Ocean, TrondheimTrondheim, Norway

F. Pohl
Marine School of Biological and Marine Sciences, University of Plymouth, Drake Circus, 4 8AA, Plymouth, UK

C. Russell
School of Geography, Geology, and the Environment, University of Leicester, Leicestershire LE1 7RH, Leicester, UK

A. Thit
Department of Science and Environment, Roskilde University, Roskilde, Denmark

Maritime Tech, Environmental and Ecotoxicological Institute, DHI, Hørsholm, Denmark

W. Umar
Leibniz Institute of Agricultural Engineering and Bioeconomy, Potsdam, Germany

D. Valero
IHE Delft Institute for Water Education, Water Resources and Ecosystems Department, Delft, The Netherlands

Karlsruhe Institute of Technology (KIT), Karlsruhe, Germany

identifies seven research goals that should be focused on in the future to enhance our knowledge on microplastics in the freshwater environment. This extended abstract presents the core message from the review paper, but for a detailed insight, please refer to the original paper [4].

15.2 Particle Properties

15.2.1 Background

Describing the physical nature of microplastic particles continues to present many difficulties [5, 6]. Currently, size, density or polymer type and shape of particles are predominantly used to describe microplastics. However, a closer look shows that this is not as straightforward as one might assumed (Table 15.1).

One of the most common definitions of microplastic is "a particle smaller than 5 mm". However, the details of how this particle **size** is determined are generally not discussed. Officially, this upper size limit for microplastics refers to the longest particle side [8]. In practice, particles can be defined by their three main dimensions

Table 15.1 Comparison of microplastic and sediment particle properties. Microplastic characteristics based on Waldschläger et al. [4], size classes for clastic sediment based on the Udden-Wentworth scale [7]

Characteristics	Microplastics	Clastic sediment
Size	0.005–5 mm	Clay: < 0.004 mm Silt: 0.004–0.063 mm Sand: 0.063–2 mm Gravel: 2–63 mm
Density	0.02–2.3 g/cm^3	2.65 g/cm^3
Shape	Pellets, fragments, foams, foils, fibers	Granular

A. Varrani
Department of Hydrology and Hydrodynamics, Institute of Geophysics, Polish Academy of Sciences, Warsaw, Poland

Department of Civil, Chemical, Environmental, and Material Engineering, University of Bologna, Bologna, Italy

A. K. Warrier
Centre for Climate Studies/Department of Civil Engineering (Manipal Institute of Technology), Manipal Academy of Higher Education, Manipal, India

L. C. Woodall
Department of Zoology, University of Oxford, Oxford OX1 3SZ, UK

N. Wu
School of Geography, Queen Mary University of London, Mile End Road, London 1 4NS, UK

(long axis, L; intermediate axis, I; short axis, S), which complicates the determination of the particle size of irregular, three-dimensional microplastic particles. Lower size limits are often related to the measurement technique, but do not indicate the longest particle side, e.g., using nets to collect water samples will influence the lower detection limit of the mean particle size. The same holds true for analytical methods used: while a sieve-based particle size is defined by the mean axis, sedimentation analyses depend on both the longest and the mean (e.g., largest projected surface area) or on all dimensions [9]. Manual size measurements are highly subjective, while imaging techniques capture only two of the three particle sides and therefore do not provide a comprehensive picture either [5, 10]. In sediment research, these particle-based descriptions are rarely used; instead, mass-based values (D_{50}, D_{90}) are applied [11, 12]. This is much more straightforward for clastic sediments, since they are mainly considered as particle mixtures of equal density. Particle size distributions based on sieve analyses have not been used for microplastics because the different particle densities complicate a mass-based analysis and the general interest has so far been on individual particles or particle counts [13].

Determining the **density** of microplastics is also not straightforward, as the particles can exhibit three different densities: (1) the initial polymer density, (2) the environmental density, which includes biofouling, aggregation with environmental substances [14] and fragmentation/degradation [15] and 3) the density determined in the lab after cleaning off the biofouling and the environmental substances [16]. Although this has also been observed for natural sediments, the influence on density and especially transport processes does not seem to be as pronounced there, as the original density of sediments (~2.65 g/cm^3 [17]) is considerably heavier than the density of water, so that a change in density has less impact [18]. Which density is most useful for describing microplastics will need to be discussed in the future. However, we must be aware that determining the polymer type of a particles alone does not determine the particle density.

Over the past 20 years, several **shape** categories for microplastics have evolved: pellets, fragments, fibres, foams and films [5]. These categories can be used to identify the origin of particles by dividing them into primary (pellets) and secondary (fragments, fibres, foams and films) microplastics, but they also pose some problems: Since these categories are not based on geometric standards, it is difficult to compare different studies [19]. One study might label a particle as foam, another as a fragment. Moreover, these shape categories are not practical for modelling the transport behaviour of these particles, as they are not parameterized and thus cannot be implemented into models. With this problem in mind, a look at sediment research may be helpful. Sediments are described using either geometric standards (e.g., ellipsoid, cylinder) [20] or shape descriptors that use the three main dimensions of the particles and their ratio (e.g., Corey shape factor) or at least a standardized matrix to categorize shape (e.g., flatness, roundness) [10, 21]. Whether these shape descriptors can also be used for microplastics remains to be investigated in detail, but measuring the three principal dimensions seems to be a first step to improve the comparability of different studies [22, 23].

15.2.2 Research Goals and Tasks

Based on the presented shortcomings that currently still exist in the description of microplastic particles, the following research objective was defined.

RG1: To improve and standardize descriptions of microplastic particles

Achieving this goal will improve the comparability of individual studies and allow us to understand the effects of particle properties on transport and particle ecotoxicology. To reach this goal, four main tasks need to be accomplished: The determination of particle size and main dimensions of non-spherical microplastics needs to be improved, e.g., by using shape descriptors developed for clastic sediments. The suitability of the currently used shape categories (pellets, fragments, films, fibres) needs to be evaluated and, if necessary, new shape categories need to be developed. The impact of different shapes and the deformability of microplastic particles on their transport behaviour needs to be studied in more detail, and the implications of the density of microplastics being close to that of water and its changes in the environment need to be evaluated.

15.3 Transport Processes

15.3.1 Background

Historically, transport modelling has often considered microplastics as artificial sediment by applying transport descriptions originally developed for natural sediments to microplastics [24]. This transfer from sediment research, which in contrast to microplastics research has been conducted for many decades (e.g., [1, 2]), may provide important insights for our understanding of microplastics. However, it needs to be examined whether this transfer leads to appropriate results (Fig. 15.1).

In the fluvial environment, microplastics can either be transported downstream or deposit due to sedimentation. Transport can occur as surface, suspended or bed load, with surface load being a new transport mode compared to clastic sediment. The diversity of particle properties leads to greater difficulty in describing the vertical distribution of microplastics in the water column – which has been studied theoretically [25] and with few field studies [26–28], but not with systematic studies on the influence of different environmental conditions on the vertical distribution. An important factor affecting the vertical distribution are the settling and rising velocities, which have often been calculated in the past based on sedimentological theory. However, it became clear that these calculations are not suitable for microplastics due to their asymmetric shapes [29–31]. Important influences on vertical velocities that are not yet fully understood include biofouling, degradation/fragmentation, and particle aggregation with natural substances—all processes that occur for natural

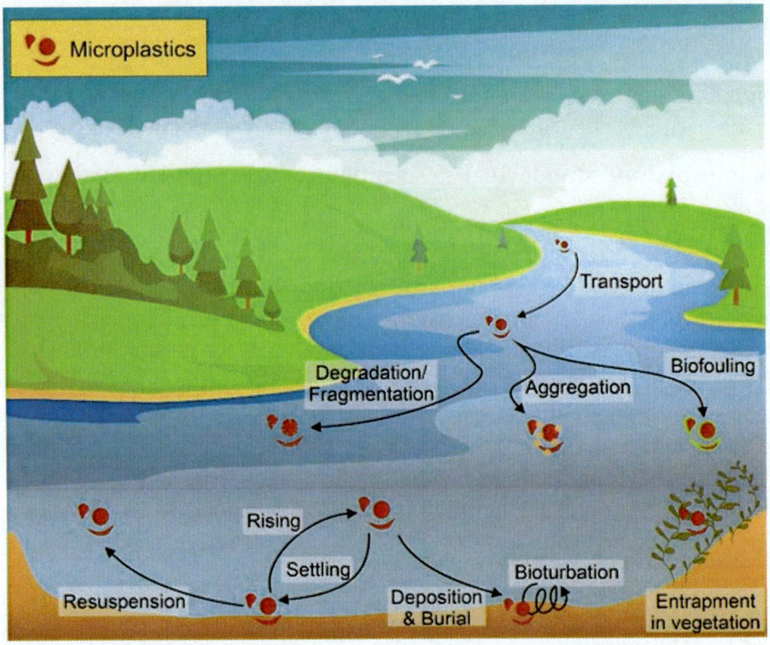

Fig. 15.1 Transport processes of microplastics in the fluvial environment

sediment too, but that are not as important due to the higher particle densities of sediments in comparison to microplastics. As for downstream transport, resuspension thresholds are highly important, which are also not well described by using calculations from sediment transport (e.g. Shields Diagram) [32]. The resuspension is additionally influenced by the burial and bioturbation of microplastics, which points to another topic that remains understudied in the microplastic context: the influence of flora (e.g. entrapment in vegetation) and fauna (e.g. ingestion and excretion) on transport behaviour [33, 34]. These shortcomings in our current understanding of microplastic transport are also relevant to numerical fate modelling. While zero to threedimensional models are available at various scales, appropriate parameterizations for various processes (e.g., particle properties, settling/rising velocities, resuspension thresholds, aggregation, biofouling, fragmentation and beaching) are still lacking.

15.3.2 Research Goals and Tasks

We have identified four transport-related research objectives that need to be addressed. Achieving these goals will enable better implementation of time-dependent changes in numerical models (RG 2) and improve our understanding of microplastic transport under different influences (RG 2, 3, 4, 5), as well as the reliability of environmental monitoring (RG 3).

RG2: To understand and quantify time variable particle property changes and interactions with other environmental substances and their impact on microplastic transport
RG3: To evaluate the vertical distribution of aquatic microplastic and their transport in the water column
RG4: To evaluate differences in the erosion and deposition behaviour between microplastics and sediments
RG5: To understand and quantify the impact of biota on microplastic transport

To address these research goals, several tasks need to be performed. This includes investigating time-dependent changes in particle properties, particularly due to biofouling, degradation/fragmentation, and aggregation, as well as parameterizing their influence on transport behaviour. Based on the concepts of sediment transport, the distribution of the total microplastic transport as surface, suspended, and bedload transport needs to be quantified, and the mechanisms of bedload transport of microplastics described. The influence of particle properties (e.g., shape, surface properties) on the onset of motion needs to parameterized and the resuspension of aggregated particles, both for microplastics-microplastics and microplastics-sediment mixtures needs to be determined. Finally, vegetation-induced sedimentation of microplastics needs to be compared to that of natural sediments, and the effects of biota on resuspension thresholds and bedload transport need to be quantified.

15.4 Monitoring Methods

15.4.1 Background

While the initial focus was primarily on the marine environment, the fluvial environment appears to be as heavily polluted as the oceans and may well be considered a temporary sink of microplastics [35, 36]. Despite this, the Water Framework Directive (WFD) adopted by the European Union in 2000 to assess the ecological status of watercourses does not include pollution limits or monitoring requirements for microplastics. This may be due to the lack of studies on microplastic concentrations in rivers prior to 2010 [35], as, in comparison, the 2008 Marine Strategy Framework Directive already lists plastic as a hazard and calls for a reduction in emissions to the oceans. It is therefore probably only a matter of time before microplastics are included in the monitoring framework of the WFD.

However, representative monitoring would hardly be possible at present due to the lack of a sound sampling concept. Representative sampling in the fluvial environment remains difficult due to dynamic flow characteristics, spatial and temporal limitations in sampling, neglect of the influence of hydrology, consideration of only one environmental compartment (e.g., surface water or sediment), and size- or shape-selective sampling [4]. As a result, it is difficult to estimate the concentration throughout the water column [25, 37].

Regarding sampling methods for the benthic environment, we often rely on techniques from sediment sampling, such as Ponar or Van Veen grabs, coring, trapping and drilling [38–40]. In the aquatic phase, one has to differentiate between surface samples, samples of suspended particles in the water column or samples of the bedload transport. For surface sampling, we cannot refer to sediment techniques. However, for sampling suspended transport, sediment traps [41–44] and Van Dorn samplers [45] can be transferred to microplastic sampling. For more advanced sampling as well as identification techniques, the reader is hereby referred to the original paper [4] due to space limitations in this abstract.

15.4.2 Research Goals and Tasks

With respect to field monitoring techniques, we have identified a research objective that will provide representative descriptions of the occurrence and types of microplastics in the fluvial environment and improve the comparability of results from different studies:

RG6: To improve and standardize sampling methods

To achieve this goal, we need to develop refined sampling techniques and parameters for sample collection, preparation and analysis, taking into account the transport mechanisms of microplastics. Furthermore, the application of sediment sampling methods such as flow-integrated sampling or the development of innovative techniques such as remote sensing, hyperspectral imaging, acoustics or laser diffraction methods for microplastics research must be explored.

15.5 Ecotoxicology

15.5.1 Background

In terms of ecotoxicology, the main difference between microplastics and clastic sediments is probably their composition: while sediments are inert, microplastics may contain toxic chemical contaminants (e.g., additives and by-products of their degradation) [46], which can leach in the aquatic environment or after ingestion [47]. In contrast to the chemical toxicity, toxicity due to their physical properties and the comparison to the ecotoxicity of natural substances has only recently been studied [48–50]. Additionally, the influence of particle shape and diameter on their ecotoxicity has not yet been sufficiently explored. For a more detailed analysis of the current state of knowledge on this topic, the reader is referred to the original review paper [4].

15.5.2 Research Goals and Tasks

In addition to the potential harm due to chemical contamination of microplastics, the physical impact of these particles needs to be studies in comparison to the effect of naturally occurring particles such as sediment. This led to the following research goal:

RG7: To study the drivers of microplastic toxicity in comparison with sediment particles

Investigating these factors will improve our knowledge of the relative importance of the polymers themselves, the plastic-associated chemicals, and the physical nature of the particles to their ecotoxicity compared to other particles in the environment. To achieve this goal, experimental design and toxicity tests must be improved by using natural particles similar to the microplastics under study as reference materials.

15.6 Conclusions

It is often assumed that the paradigms used in sediment research are also applicable to describe the transport of microplastics, as they can be considered as a type of artificial sediment that behaves similarly in the aquatic environment. However, microplastic particles are diverse and characterized by highly variable densities, particle sizes, and shapes, which distinguishes them from natural sediments. Therefore, our descriptions of microplastic particles need to be improved and standardized to gain insight into how particle properties affect the transport of microplastics in the environment and to improve the comparability of different studies. Particle description methods from sedimentology, such as the use of shape descriptors, should be further evaluated for microplastics research. In addition to the high variability of microplastic particles, environmental changes in particle properties, such as biofouling, aggregation or fragmentation, complicate the understanding of their behaviour and particle description. These processes, which have received little attention in the sediment field, could be significant for microplastic transport and require further research.

The classification of sediment transport into suspended sediment and bedload transport has already been applied to microplastics research and extended to include a third class, surface transport. For bedload transport, there are still many unanswered research questions, such as the flow conditions that lead to particle movement near the bottom, the amounts transported, and the interactions of particles with the sediment bed. Shifts between these transport modes, usually estimated from particle size and density of natural sediments, are still poorly understood for microplastics and require the development of metrics that relate particle properties and hydrology to transport mode. Particle properties and transport processes are important aspects to consider when developing representative monitoring strategies for defining microplastic concentrations. Detailed knowledge about environmental concentrations is essential to validate transport models, to use environmentally relevant concentrations and particle properties in toxicological assessments and to develop targeted mitigation strategies. In the future, ecotoxicology should not only focus on the chemical effects of microplastics, but also on particle-related effects, which must be evaluated in comparison to naturally occurring particles.

To conclude, even after 20 years, microplastics research remains a comparatively young research field that has many open questions to answer, but can certainly learn from other disciplines and thus accelerate progress significantly

References

1. Wadell H (1933) Sphericity and roundness of rock particles. J Geol 41:310–331
2. Shields A 1936 Anwendung der Ähnlichkeitsmechanik und der Turbulenzforschung auf die Geschiebebewegung. Doctoral Thesis: Berlin
3. Rouse H (1937) Modern conceptions of the mechanics or fluid turbulence. Trans Am Soc Civ Eng 102:463–505
4. Waldschläger K, Brückner MZ, Carney Almroth B, Hackney CR, Adyel TM, Alimi OS, Belontz SL, Cowger W, Doyle D, Gray A et al. (2022) Learning from natural sediments to tackle microplastics challenges: A multidisciplinary perspective. Earth-Sci Rev, 228, 104021
5. Rochman CM, Brookson C, Bikker J, Djuric N, Earn A, Bucci K, Athey S, Huntington A, McIlwraith H, Munno K et al (2019) Rethinking microplastics as a diverse contaminant suite. Environ Toxicol Chem 38:703–711
6. Hartmann NB, Hüffer T, Thompson RC, Hassellöv M, Verschoor A, Daugaard AE, Rist S, Karlsson T, Brennholt N, Cole M et al (2019) Are we speaking the same language? recommendations for a definition and categorization framework for plastic debris. Environ Sci Technol 53:1039–1047
7. Wentworth CK (1933) The shapes of rock particles: a discussion. J Geol 41:306–309
8. Arthur C, Baker J, Bamford H (2009) Proceedings of the International Research Workshop on the Occurrence, Effects and Fate of Microplastic Marine Debris. Sept 9–11:2008
9. Church MJ, Grain size and shape (2003) In Sedimentology: Springer Netherlands: Dordrecht, pp 544–554
10. Blott S, PYE K (2006) Particle size distribution analysis of sand-sized particles by laser diffraction: an experimental investigation of instrument sensitivity and the effects of particle shape: particle size distribution analysis of sands by laser difraction. Sedimentology, 671– 685
11. Tanner WF (1995) Environmental clastic granulometry. Florida Geological Survey. Special Publication 40. 163
12. Boggs S, JR. (2009) Petrology of sedimentary rocks, cambridge university press: Cambridge
13. Kooi M, Koelmans AA (2019) Simplifying microplastic via continuous probability distributions for size, shape, and Density. Environ Sci Technol Lett 6:551–557
14. Leiser R, Wu G-M, Neu TR, Wendt-Potthoff K (2020) Biofouling, metal sorption and aggregation are related to sinking of microplastics in a stratified reservoir. Water research, 176, 115748

15. Skalska K, Ockelford A, Ebdon JE, Cundy AB (2020) Riverine microplastics: Behaviour, spatiotemporal variability, and recommendations for standardised sampling and monitoring. J Water Process Eng 38:101600
16. Lechthaler S, Stauch G, Schüttrumpf H (2020) Oil extraction as separation method for microplastic in sediment samples. In Proceedings of the 2nd International Conference on Microplastic Pollution in the Mediterranean Sea; Cocca M, Di Pace E, Errico ME, Gentile G, Montarsolo A, Mossotti R, Avella M, Eds.: Springer Int Publ: Cham, pp 282–286
17. Chamley H (1990) Sedimentology, Springer Berlin Heidelberg: Berlin, Heidelberg
18. Fang HW, Lai HJ, Cheng W, Huang L, He GJ (2017) Modeling sediment transport with an integrated view of the biofilm effects. Water Resour Res 53:7536–7557
19. Kooi M, Primpke S, Mintenig SM, Lorenz C, Gerdts G, Koelmans AA (2021) Characterizing the multidimensionality of microplastics across environmental compartments. Water Res 202:117429
20. Le Roux JP (2005) Grains in motion: A review. Sediment Geol 178:285–313
21. Corey (1949) Influence of shape on the fall velocity of sand grains
22. van Melkebeke M, Janssen C, de Meester S (2020) Characteristics and sinking behavior of typical microplastics including the potential effect of biofouling: implications for remediation. Environ Sci Technol 54:8668–8680
23. Cowger W, Gray A, Christiansen SH, DeFrond H, Deshpande AD, Hemabessiere L, Lee E, Mill L, Munno K, Ossmann BE et al (2020) Critical review of processing and classification techniques for images and spectra in microplastic research. Appl Spectrosc 74:989–1010
24. Kooi M, Besseling E Kroeze C, van Wenzel AP, Koelmans AA, van Wezel AP (2018) Modeling the fate and transport of plastic debris in freshwaters: review and guidance. In Freshwater Microplastics: Emerging Environmental Contaminants*?*; Wagner M, Lambert S, Eds.: Springer: Cham, 58, pp 125–152
25. Cowger W, Gray AB, Guilinger JJ, Fong B, Waldschläger K (2021) Concentration depth profiles of microplastic particles in river flow and implications for surface sampling. Environ Sci Technol 55:6032–6041
26. Lenaker PL, Baldwin AK, Corsi SR, Mason SA, Reneau PC, Scott JW (2019) Vertical distribution of microplastics in the water column and surficial sediment from the milwaukee river basin to lake michigan. Environ Sci Technol 53:12227–12237
27. Morritt D, Stefanoudis PV, Pearce D, Crimmen OA, Clark PF (2014) Plastic in the Thames: a river runs through it. Mar Pollut Bull 78:196–200
28. Haberstroh CJ, Arias ME, Yin Z, Wang MC (2020) Effects of hydrodynamics on the crosssectional distribution and transport of plastic in an urban coastal river. Water Environ Res: Res Publ Water Environ Fed
29. Waldschläger K, Schüttrumpf H (2019) Effects of particle properties on the settling and rise velocities of microplastics in freshwater under laboratory conditions. Environ Sci Technol 53:1958–2066
30. Waldschläger K, Born M, Cowger W, Gray A, Schüttrumpf H (2020) Settling and rising velocities of environmentally weathered micro—and macroplastic particles. Environ Res 110192
31. Kuizenga B, van Emmerik T, Waldschläger K, Kooi M (2022) Will it float? rising and settling velocities of common macroplastic foils. ACS ES&T water 2:975–981
32. Waldschläger K, Schüttrumpf H (2019) Erosion behavior of different microplastic particles in comparison to natural sediments. Environ Sci Technol 53:13219–13227
33. Coppock RL, Lindeque PK, Cole M, Galloway TS, Näkki P, Birgani H, Richards S, Queirós AM (2021) Benthic fauna contribute to microplastic sequestration in coastal sediments. J Hazard Mater 415:125583
34. Zhang H (2017) Transport of microplastics in coastal seas. Estuar Coast Shelf Sci 199:74–86
35. Waldschläger K, Lechthaler S, Stauch G, Schüttrumpf H (2020) The way of microplastic through the environment—Application of the source-pathway-receptor model (review). The way of microplastic through the environment - Application of the source-pathway-receptor model (review). Sci Total Environ 713:136584

36. van Emmerik T, Mellink Y, Hauk R, Waldschläger K, Schreyers L (2022) Rivers as plastic reservoirs. Front. Water 3
37. Liedermann M, Gmeiner P, Pessenlehner S, Haimann M, Hohenblum P, Habersack H (2018) A methodology for measuring microplastic transport in large or medium rivers. Water 10:414
38. Storlazzi CD, Field ME, Bothner MH (2011) The use (and misuse) of sediment traps in coral reef environments: theory, observations, and suggested protocols. Coral Reefs 30:23–38
39. Tidjani AD, Bielders CL, Rosillon D, Ambouta KJ-M (2011) Uncertainties in Plot-Scale mass balance measurements using aeolian sediment traps. Soil Sci Soc Am J 75:708–718
40. Kondolf GM, Lisle TE, Wolman GM (2003) Bed sediment measurement. In Tools in Geomorphology, Kondolf GM, Piegay H, Eds.: John Wiley and Sons Ltd: New York, NY, USA
41. Lorenz C, Roscher L, Meyer MS, Hildebrandt L, Prume J, Löder MGJ, Primpke S, Gerdts G (2019) Spatial distribution of microplastics in sediments and surface waters of the southern North Sea. Environmental pollution (Barking, Essex : 1987), 252, 1719–1729
42. Bagheri T, Gholizadeh M, Abarghouei S, Zakeri M, Hedayati A, Rabaniha M, Aghaeimoghadam A, Hafezieh M (2020) Microplastics distribution, abundance and composition in sediment, fishes and benthic organisms of the Gorgan Bay. Caspian sea. Chemosphere 257:127201
43. Fraser MA, Chen L, Ashar M, Huang W, Zeng J, Zhang C, Zhang D (2020) Occurrence and distribution of microplastics and polychlorinated biphenyls in sediments from the Qiantang River and Hangzhou Bay. China Ecotoxicol Environ Saf 196:110536
44. Rios Mendoza LM, Leon Vargas D, Balcer M (2021) Microplastics occurrence and fate in the environment. Curr Opin Green Sustain Chem 32:100523
45. Wren DG, Barkdoll BD, Kuhnle RA, Derrow RW (2000) Field techniques for suspendedsediment measurement. J Hydraul Eng 126:97–104
46. Wiesinger H, Wang Z, Hellweg S (2021) Deep dive into plastic monomers, additives, and processing aids. Environ Sci Technol 55:9339–9351
47. Hermabessiere L, Dehaut A, Paul-Pont I, Lacroix C, Jezequel R, Soudant P, Duflos G (2017) Occurrence and effects of plastic additives on marine environments and organisms: A review. Chemosphere 182:781–793
48. Schür C, Zipp S, Thalau T, Wagner M (2020) Microplastics but not natural particles induce multigenerational effects in Daphnia magna. Environ Pollut 260:113904
49. Scherer C, Weber A, Stock F, Vurusic S, Egerci H, Kochleus C, Arendt N, Foeldi C, Dierkes G, Wagner M et al (2020) Comparative assessment of microplastics in water and sediment of a large European river. Sci Total Environ 738:139866
50. Zimmermann L, Göttlich S, Oehlmann J, Wagner M, Völker C (2020) What are the drivers of microplastic toxicity? Comparing the toxicity of plastic chemicals and particles to Daphnia magna. Environ Pollut 267:115392

Chapter 16
Environmental Impact of Innovative Microplastics Filtration System for Washing Machines & Marinas

Hakim El Khiar, Kostja Klabjan, Jan Puhar, and Annamaria Vujanović

16.1 Introduction

The Mediterranean coastal zone is a densely populated, high activity region, where its topography and lack of efficient waste management in many countries have led to the accumulation of plastic debris in the Mediterranean Sea. Published studies report that plastic dominates in the marine litter [1], where 83% of plastic items in samples collected are microplastics [2].

Studies show that 36% of global microplastic pollution comes from washing clothes [3] while over 9% comes from ship-based losses [4]. CleraOne is a sustainable wastewater treatment company whose chemicals-free water recycling system enables the reuse of wastewater and stops the discharge of microplastics. Its innovative membrane allows a high-water permeation and flow while maintaining low pressure, with membrane pores only 0.01 microns in size. The purified effluent water is organoleptically cleaner than tap water and can be reused in all washing processes for industrial purposes. Instead of common linearly aligned filters with higher energy and spatial footprints, the proposed system provides a synergistic, automated innovation that enables the removal of microparticles ranging in size from 5 mm to 0.10 microns. The water recycling system can be retrofitted to reuse wastewater in all laundry systems, vessels washers, and cruise systems. CleraOne's device shows to retain 99% of microplastic and enables 70% wastewater reuse.

As the main source of the world's microplastic pollution, laundry operations are key hotspots for analysis in this field, and the need for commercial laundries to

H. E. Khiar · K. Klabjan
Energ+ d.o.o., Ferrarska ulica 30, 6000 Koper, Slovenia

J. Puhar · A. Vujanović (✉)
Faculty of Chemistry and Chemical Engineering, University of Maribor, Smetanova 17, 2000 Maribor, Slovenia
e-mail: annamaria.vujanovic@um.si

M. Cocca et al. (eds.), *Proceedings of the 3rd International Conference on Microplastic Pollution in the Mediterranean Sea*, Springer Water,
https://doi.org/10.1007/978-3-031-34455-8_17

increase their sustainability has been clearly stated [5]. Thus far, the environmental benefits of CleraOne's groundbreaking filtration system have not been analysed in full. In this work, environmental impacts are quantified using the Life Cycle Assessment (LCA) methodology, which is performed using OpenLCA software [6] with the integrated database Ecoinvent 3.6 [7].

16.2 Case Study Framework

The analysis in this study considers the wastewater treatment for an average commercial laundry's full operational year. To consider an average laundry operation and its wastewater treatment, it is assumed that a commercial washing machine uses 34.7 thousand gallons [8] or 157.9 m^3 of water per year which requires treatment. To quantify the benefits of using CleraOne's filtration system over other similar practices, a reverse osmosis treatment process is assumed, as this is the methodology applied in CleraOne's device as well as an increasingly popular choice in household as well as commercial laundry systems. All relevant material and energy inputs required for treatment by reverse osmosis are applied to the LCA.

Reverse osmosis water filter systems are generally comprised of several parts, including (a) a carbon polishing filter, (b) a granular activated carbon filter, (c) a sediment pre-filter, and (d) an ultrafine reverse osmosis membrane. It is recommended by most reverse osmosis system providers that all filter parts except the membrane are replaced once per 6,000 gallons (or 27.3 m^3) of discharged fresh water [9]. The membrane itself lasts much longer and its replacement is therefore excluded from this study.

The material inputs considered in this LCA study are based on the reverse osmosis system filters which require frequent replacement. Firstly, the carbon polishing filter is made from activated carbon and is sold at an average weight of 1.13 kg per unit [10]. The granular activated carbon filter utilizes carbon granules and has an average weight of 2.27 kg per unit [10]. Finally, the sediment pre-filter utilizes polypropylene film and weighs on average 0.17 kg per unit [10]. By considering filter replacement frequency at one of each filter unit per 6,000 gallons, the mass amounts of newly produced filter material in a full year of laundry operations were calculated and used as input data for the environmental assessment.

Additionally, reverse osmosis wastewater treatment requires energy input in the form of electricity, where an average of about 6.5 kWh per m^3 of produced freshwater [11] is used. The majority of the energy at over 84% is required during the high-pressure water pumping phase, with the remaining energy mainly used for final water product transfer and supply, initial water supply, and pre-treatment [12]. To consider electricity, the Slovenian electricity mix, which derives around a third of its energy from lignite, a third from nuclear power, and a third from hydropower [13], was used in this study. Table 16.1 presents inventory data for the LCA study case, comprised of material and energy input amounts for the functional unit of 157.9 m^3 of treated water during a full operating year.

Table 16.1 Inventory data with material and energy inputs for the reverse osmosis treatment process

Parameter	Amount	Unit
Activated carbon	6.55	kg
Carbon granules	13.15	kg
Polypropylene film	0.98	kg
Electricity	1026.55	kWh

The reverse osmosis treatment process' ability to remove microplastics from wastewater is additionally considered. In this case it is assumed that a commercial laundry machine utilizes on average 0.03 m^3 of freshwater per 1 kg of laundry [14]. Due to lack of data on average laundry load microplastic release, the problem is simplified by considering average microplastic release during polyester shirt washing. According to a study on microfiber release from polyesters, on average 0.01% of a shirt's mass is released during each wash [15]. By considering the weight of an average medium size shirt at 0.138 kg, the total amount of microplastic discharged into the environment without wastewater treatment would amount to 3.8 kg during the operational year. By utilizing either reverse osmosis treatment or CleraOne's treatment device, around 99% of microplastic is retained and prevented from release into water and soil.

16.3 Results and Discussion

Results for the full LCA of one laundry operational year using the reverse osmosis wastewater treatment method are presented in this section. All considered impact categories reflect the impact produced throughout the life cycle of a product or service, in the quantity required to reach the functional unit. The potential for abiotic depletion represents the lack of resources such as minerals and clay, while abiotic depletion of fossil fuels includes fuels such as coal. Acidification expresses the impact of total emissions of acid gases such as sulphur oxides and nitrogen oxides, which can acidify and negatively impact rivers, lakes and soil. Eutrophication expresses the excessive growth of algae due to the release of nutrients containing phosphorus and nitrogen into water, causing biomass growth. Human toxicity potential and ecotoxicity potentials for fresh and marine wasters and soil are determined by chemically specific characteristic factors that quantify the impact of chemical emissions on human health and the environment by assessing exposure and impact. Global warming potential (GWP), also referred to as greenhouse gas footprint, represents climate change over a 100-year period as total greenhouse gas emissions to the air. Ozone depletion represents the relative impact of total emissions of stratospheric ozone-depleting gases, and photochemical oxidation reflects total emissions of volatile organic compounds and nitrogen oxides that form smog, or photochemical ozone. All considered impact categories are measured in the units as shown in Table 16.2.

Table 16.2 Impact category results

Parameter	Amount (electricity)	Amount (activated carbon)	Amount (carbon granules)	Amount (polypropylene)	Unit
Abiotic depletion (minerals)	$7.39 \cdot 10^{-4}$	$7.55 \cdot 10^{-5}$	$5.78 \cdot 10^{-5}$	$5.82 \cdot 10^{-5}$	kg Sb eq
Abiotic depletion (fossil fuels)	3,354.58	663.53	1,107.44	78.06	MJ
Acidification	8.28	0.33	0.14	$1.05 \cdot 10^{-2}$	kg SO_2 eq
Eutrophication	2.65	0.12	0.02	$3.35 \cdot 10^{-3}$	kg PO_4^{3-} eq
Freshwater ecotoxicity	446.94	21.80	1.04	3.03	kg 1,4-DB eq
Global warming	386.10	50.84	32.79	2.79	kg CO_2 eq
Human toxicity	300.62	22.78	5.21	1.12	kg 1,4-DB eq
Marine ecotoxicity	1,230,629	75,014	8,159	2,105	kg 1,4-DB eq
Ozone layer depletion	$2.68 \cdot 10^{-6}$	$1.21 \cdot 10^{-6}$	$1.43 \cdot 10^{-5}$	$9.70 \cdot 10^{-8}$	kg CFC-11 eq
Photochemical oxidation	0.32	$1.47 \cdot 10^{-2}$	$6.13 \cdot 10^{-3}$	$5.45 \cdot 10^{-4}$	kg C_2H_4 eq
Terrestrial ecotoxicity	0.64	0.06	$1.86 \cdot 10^{-2}$	$4.26 \cdot 10^{-3}$	kg 1,4-DB eq

Impact category results are presented graphically in Fig. 16.1. The dominant impact of electricity among the considered processes is notable, as it contributes a share of over 80% in the majority of all impact categories. The exceptions are abiotic depletion of fossil fuels and ozone layer depletion, which are more intensely impacted by the production of carbon granules and the related exhaustion of fossil resources. In general, activated carbon is found to have the second most harmful impact to the environment among the processes, followed by carbon granules and finally by polypropylene.

As noted by the LCA results, the impacts are divided between two process phases: electricity required to power the device, and filters requiring replacement after a certain number of washes. By utilizing CleraOne's device which automatically cleans and removes dry micropollutants from its membrane, the need for filter replacement would be excised and environmental burdens would be significantly lowered.

Results comparing the reverse osmosis wastewater treatment method with CleraOne's device regarding GWP are presented in Fig. 16.2. The figure additionally shows a scenario where the laundry water is not treated at all, which would result in dangerous amounts of microplastic being discharged into the environment. It should be noted that due to a lack of regulation on microplastic pollution in most countries, this exists as a potentially realistic scenario. By treating wastewater, the reverse

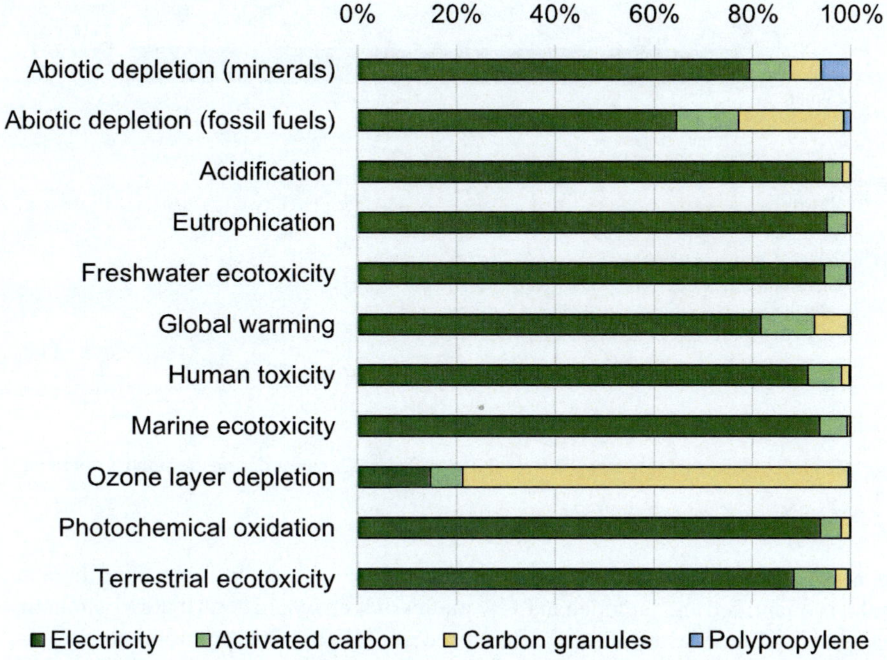

Fig. 16.1 Impact category results for the reverse osmosis method

osmosis process and the CleraOne device are shown to retain 99% of microplastics. The difference among the two processes is clearly shown in the third scenario's reduced GWP, as the CleraOne device does not require frequent replacement of carbon and polypropylene-based filters.

16.4 Conclusions

This study on microplastic filtration systems presents the environmental benefits gained by utilization of CleraOne's innovative wastewater treatment device, requiring no filter replacements and operating with lower energy consumption and consequently a lower environmental burdens. While electricity consumption is shown to be the main contributor to the assessed environmental impact categories, it should be noted that this factor is dependent on the electricity source used, thus the impacts of wastewater treatment could be further mitigated by opting for electricity derived from renewable sources.

A limitation of this study is its lack of consideration of microplastic pollution within the LCA, which has been a major scientific challenge as this relatively recent pollution issue has not been sufficiently characterized by traditional LCA databases and impact assessment methods [16]. In future work, it is imperative that the specific

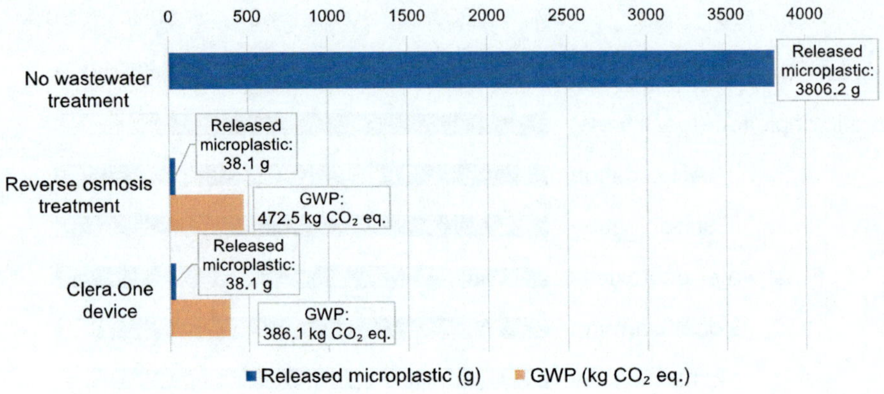

Fig. 16.2 Microplastic release and GWP results for different scenarios during the laundry operating year

impacts and harms of released microplastic to both the environment and human health is quantified and included in LCA studies, which would contribute to a holistic assessment of the wastewater treatment process. Microplastic pollution characterization could be achieved similarly to examples shown in recent studies, where the amount of released microparticles was converted into pellet-equivalent point metrics on a 100-year life cycle basis [17]. Alternatively, different types of micropollutants could be converted into the impact metric of freshwater ecotoxicity potential, based on their fragmentation rate and toxicity [18].

Despite current shortcomings in the field of quantifying microplastic pollution on a life cycle level, global awareness is growing on the harmful effects of microplastic and on the imperative need for energy-efficient treatment systems. As shown in this work, the innovative CleraOne microplastic filtration system is a viable alternative to current practice, presenting an important step towards providing clean water in a sustainable way.

References

1. Munari C, Corbau C, Simeoni U, Mistri M (2016) Marine litter on Mediterranean shores: analysis of composition, spatial distribution and sources in north-western Adriatic beaches. Waste Manage 49:483–490
2. Cózar A, Sanz-Martín M, Martí E, González-Gordillo JI, Ubeda B, Gálvez JÁ, Irigoien X, Duarte CM (2015) Plastic accumulation in the Mediterranean Sea. PLoS One 10(4):e0121762
3. Armstrong M (2019) Where do the oceans' microplastics come from? https://www.statista.com/chart/17957/where-the-oceans-microplastics-come-from/. Accessed 9 April 2022
4. Armstrong M (2019) Laundry's contribution to the world's microplastic problem. https://www.statista.com/chart/17937/laundry-contribution-to-world-microplastic-problem/. Accessed 9 April 2022

5. Swartz CD, Swanepoel G, Welz PJ, Muanda C, Bonga A (2017) Water and wastewater management in the laundry industry. WRC report no. TT 703/16
6. GreenDelta (2020) openLCA V1.10.3. https://www.openlca.org. Accessed 14 June 2021
7. Ecoinvent V3.6 (2019) Ecoinvent, Technoparkstrasse 1, 8005 Zurich, Switzerland
8. National Park Service—U.S. Department of the Interior (2021) Laundry practices and water conservation. https://www.nps.gov/articles/laundry.htm. Accessed 16 July 2022
9. Bulk Reef Supply (2022) Reverse osmosis filters & media. https://www.bulkreefsupply.com/bulk-reverse-osmosis-filters-systems/reverse-osmosis-filters.html. Accessed 15 July 2022
10. Lenntech (2021) Lenntech. https://www.lenntech.com/Data-sheets/Parker_products_filtration_process-L.pdf. Accessed 17 July 2022
11. Dashtpour R, Al-Zubaidy SN (2012) Energy efficient reverse osmosis desalination process. Int J Environ Sci Dev 3(4):339
12. Gude VG (2011) Energy consumption and recovery in reverse osmosis. Desalin Water Treat 36(1–3):239–260
13. Statistical Office of the Republic of Slovenia (2021) Electricity (GWh), Slovenia, monthly. https://www.pxweb.stat.si/SiStatData/pxweb/en/Data/Data/1817601S.px/table/tableViewLayout2/. Accessed 20 September 2021
14. Danamark (2022) Optimize water and energy usage in hotel laundry machines. https://www.danamark.com/resources/improving-water-footprint-laundry-machines. Accessed 15 July 2022
15. Hernandez E, Nowack B, Mitrano DM (2017) Polyester textiles as a source of microplastics from households: a mechanistic study to understand microfiber release during washing. Environ Sci Technol 51(12):7036–7046
16. Henry B, Laitala K, Klepp IG (2019) Microfibres from apparel and home textiles: prospects for including microplastics in environmental sustainability assessment. Sci Total Environ 652:483–494
17. Saling P, Gyuzeleva L, Wittstock K, Wessolowski V, Griesshammer R (2020) Life cycle impact assessment of microplastics as one component of marine plastic debris. Int J Life Cycle Assess 25(10):2008–2026
18. Salieri B, Stoudmann N, Hischier R, Som C, Nowack B (2021) How Relevant are direct emissions of microplastics into freshwater from an LCA perspective? Sustainability 13(17):9922

Chapter 17
Towards a Microplastic-Free Ocean: Green Photocatalysis for Mitigation of Micro- and Nanoplastic Marine Pollution

Erika Iveth Cedillo-González and Cristina Siligardi

17.1 Introduction

Microplastics (MPs) are plastic debris with sizes between 0.3 μm and 5 mm [1], are present in all kinds of aquatic ecosystems and marine biota consumes them leading to dangerous health issues and even death. Due to their high area to volume ratio and their surface chemical properties, MPs adsorb persistent organic pollutants (POP's) if they are surrounded by contaminated environments [2]. Therefore, MPs can further damage the health of marine biota because they can act as POPs' vectors [2]. MPs' distribution and transfer among the four major environmental compartments (air, water, soil, and biota) [3] is probably the main reason of why MPs have been already found in humans [4, 5], making crucial to develop remediation technologies to fight MPs pollution. Since all the MPs present in aquatic ecosystems were first produced in upstream processes, a promising approach to fight them is the reduction through management before they enter aquatic ecosystems. A feasible approach to accomplish this goal is to eliminate MPs present in contaminated wastewater effluents before their discharge into waterbodies, as wastewater treatment plants (WWTPs) are a source of MPs to aquatic and terrestrial environments [2].

Photocatalysis is a water treatment process that can be adapted as tertiary treatment into WWTPs. This technology is based on the interaction of a photocatalyst with a pollutant adsorbed on its surface and light. If the photocatalyst is bombarded with

E. I. Cedillo-González (✉) · C. Siligardi
Department of Engineering "Enzo Ferrari", University of Modena and Reggio Emilia, Modena, Italy
e-mail: ecedillo@unimore.it

National Interuniversity Consortium of Materials Science and Technology (INSTM), Florence, Italy

M. Cocca et al. (eds.), *Proceedings of the 3rd International Conference on Microplastic Pollution in the Mediterranean Sea*, Springer Water,
https://doi.org/10.1007/978-3-031-34455-8_18

photons with energy equal or higher than its band gap ($E \geq E_g$), the electrons (e^-) on the valance band are transferred into the conduction band, leaving behind positive hole (h^+) (Eq. 17.1). Holes react with water or hydroxyl groups adsorbed in the surface of the photocatalyst, generating hydroxyl radicals ($OH^{\cdot}$) (Eq. 17.2). Electrons react with adsorbed oxygen to form hydroperoxyl/superoxide anion radicals ($HO_2^{\cdot}/O_2^{-\cdot}$) (Eq. 17.3). All those species are called reactive oxygen species (ROS) and promote mineralization of the organic pollutants adsorbed in the surface of the photocatalyst (Eq. 17.4).

$$Photocatalyst + hv \rightarrow h^+ + e^- \tag{17.1}$$

$$h^+ + H_2O\ (or\ OH^-) \rightarrow OH^{\cdot} \tag{17.2}$$

$$e^- + O_2 \rightarrow HO_2^{\cdot}/O_2^{-\cdot} \tag{17.3}$$

$$Pollutant + ROS \rightarrow CO_2 + H_2O \tag{17.4}$$

By photocatalysis, MPs can be mineralized to CO_2 and H_2O or, be degraded into less toxic substances. However, MPs are a very particular kind of pollutants, with characteristics that make degradation challenging.

Here, we report the visible light photocatalysis of different MPs: polyethylene (PE), polystyrene (PS) and polyethylene terephthalate (PET) of different sizes. Experiments were conducted at several pH and temperature values using visible-light photocatalysts (N-TiO_2, C,N-TiO_2, C,N-TiO_2/SiO_2) synthetized by a green synthesis route using renewable raw materials as doping source. Furthermore, to evaluate the real use of this technology in WWTPs, experiments were performed in aqueous medium.

17.2 Experimental

17.2.1 Photocatalysts Synthesis and Characterization

N-TiO_2 and C,N-TiO_2 photocatalysts were prepared by the synthesis reported by Zheng et al. [6], using proteins from the extrapallial fluid (EPF, [Proteins] = 10 ppm) of *Mytilus Edulis* mussels as renewable source of N and C. For C,N-TiO_2/SiO_2 photocatalyst, tetraethyl orthosilicate, isopropyl alcohol, distilled water and concentrated nitric acid were added to the synthesis. The crystalline composition of the samples was determined by X-ray diffraction (XRD), using a Siemens D5000 diffractometer. Carbon and nitrogen doping was determined by elemental analysis by a CHNS/O Analyzer 2400 Perkin Elmer Series II instrument. The band gap (E_g) values were calculated from their diffuse reflectance spectra (DRS) using the Kubelka–Munk theory. Measures were carried out on a Jasco V-670 UV–Vis/NIR spectrophotometer equipped with an ILN-725 integration sphere. The surface area was estimated by nitrogen adsorption using a Micromeritics TriStar II Plus 3.01 surface area analyzer. Microstructural observation was carried out by Field Emission Scanning Electron Microscopy, using a FEI Nova NanoSEM 450 Field Emission SEM.

17.2.2 Film Preparation

Glass substrates were coated with a 10 wt/v % aqueous dispersion of N-TiO_2 using a CGV USB Dip-coater with an immersion and emersion rate of 100 mm/min.

17.2.3 **Microplastics Obtainment and Characterization**

Primary PS MPs and nanoplastics (NPs, particle size $\leq 0.3\ \mu m$) were prepared by the surfactant-free emulsion polymerization synthesis reported in Ref. [7]. The reaction was conducted for 10 min or 20 h to obtain PS NPs or MPs, respectively. Primary PE MPs were obtained from a commercial scrub using the extraction methodology of Napper et al. [8]. Secondary PET MPs were obtained by grinding a PET food container with a coffee grinder and sieving to get MPs with sizes $\leq 500\ \mu m$. The particle size of all MPs and NPs was estimated by SEM, using an Environmental SEM FEI-Quanta 200. Polymer type was analysed by Fourier-transformed infrared spectroscopy (FTIR) using a BRUKER Alpha II spectrometer equipped with a diamond ATR accessory.

17.2.4 Photocatalytic Experiments

17.2.4.1 Photocatalysis of Primary PS NPs and MPs

A batch-type reactor was filled with 50 mL of a 1.8 wt. % aqueous dispersion of the NPs or 50 mL of a 0.01 wt. % aqueous dispersion of the MPs. The N-TiO_2 films were placed inside the dispersion. The reaction was performed in a chamber equipped with a 50 W IP65 LED Artlite visible lamp (400 – 800 nm). The experiments were performed for 300 min at T_{room} and 300 rpm. The photolysis tests were carried out at the same conditions, but using a N-TiO_2-free glass substrate. Degradation was followed by turbidimetry (Lovibond TB 250 WL turbidimeter).

17.2.4.2 Photocatalysis of Primary PE MPs and Secondary PET MPs

The effect of three different temperature and pH values on the PE MPs degradation was evaluated using the *Design of Experiments* (DoE) approach. Experiments were conducted in a batch-type reactor, using 50 mL of a 0.4 wt/vol % PE MPs dispersion and C,N-TiO_2 powders. A reaction chamber equipped with a temperature bath (±2 °C) and a 50 W IP65 LED Artlite visible lamp set at 57.2 ± 0.3 W/m^2 was used (50 h). Photocatalytic degradation of PET MPs was done for 120 h using the same procedure but at T_{room} and two pH values (6 and 8). MPs degradation was followed by gravimetry.

17.3 Results and Discussion

17.3.1 Photocatalysis of Primary PS NPs and MPs

In the photocatalytic degradation process, adsorption of the pollutant on the surface of the photocatalyst is critical for degradation. A challenge in the photocatalytic degradation of MPs is to deal with their almost null adsorption on the photocatalyst's surface due to their large particle size compared with the sizes of common nanoparticle photocatalysts. Thus, the purpose of this part of the research was to determine whether a photocatalyst in the form of films (the most preferred form for photocatalysts to be used in tertiary treatments in WWTPs) is able to degrade PS MPs and NPs in aqueous media.

Table 17.1 presents the properties of the N-TiO_2 photocatalyst, which were found to be suitable for performing visible light photocatalysis of aquatic pollutants. It was found that this sample is mainly composed of anatase TiO_2, although traces of brookite TiO_2 were also detected. The sample is composed by agglomerated particles of 3 μm of diameter, which are in turn formed by small particles of ~10 nm. Nitrogen adsorption revealed a high value of surface area and the presence of mesoporosity. Finally, the incorporation of N (from the mussel's proteins) promoted a E_g value suitable for visible light absorption.

Figure 17.1a, b show the SEM micrographs of the PS MPs and NPs, respectively. PS NPs have sizes of 278.6 ± 9.1 nm while PS MPs present sizes of 1.04 ± 0.03 μm. Figure 17.1c shows that both materials are composed of PS, as its characteristic absorption bands are present in both FTIR spectra. Figure 17.1d presents the results of the photocatalytic experiments. For both sizes, photolysis resulted in a limited reduction of the concentration of PS plastic. The trends of the red and green curves of Fig. 17.1d indicate that if the reaction time is increased, further reductions of the concentration of PS are hard to achieve by photolysis. On the other hand, the black and blue curves of Fig. 17.1d show that photocatalytic oxidation (PCO) thanks to the presence of the N-TiO_2 film in the reaction media promotes a greater reduction of the PS concentration. This reduction was attributed to the formation of ROS and

Table 17.1 Properties of the N-TiO_2 photocatalyst

Property	N-TiO_2	C,N-TiO_2	C,N-TiO_2/SiO_2
Crystalline composition	Anatase TiO_2 (main), brookite TiO_2	Anatase TiO_2 (main), rutile TiO_2	Anatase TiO_2
Particle size, nm	Agglomerates of 3 μm, formed by particles of 10 nm	Agglomerates of 3 μm, formed by particles of 10 nm	Agglomerates of 3 μm, formed by particles of 20 nm
Surface area, m^2/g	193	219	313
Type of porosity	Mesoporous	Mesoporous	Mesoporous
E_g, eV (wavelength of absorption, nm)	3.0 (414)	2.9 (388)	2.24 (553)
Nitrogen content, wt. %	0.16	0.41	0.36
Carbon content, wt. %	–	1.31	3.27

the interaction of such species with the PS particles. Interestingly, it was found that the particle size of the PS debris has not an influence on degradation. For both MPs and NPs, the presence of N-TiO_2 increases in a similar degree the removal of PS (removal increases 3.7-times for MPs and 3.4-times for NPs). These results confirm the hypothesis that the interaction of photocatalytic nanoparticles with MPs pollutants is a challenge that should be solved to apply this technology for the removal of marine MPs pollution.

17.3.2 Photocatalysis of Primary PE MPs and Secondary PET MPs

Table 17.1 presents the properties of the C,N-TiO_2 and C,N-TiO_2/SiO_2 photocatalysts. Both materials are mainly composed of anatase TiO_2, although traces of rutile were detected in the C,N-TiO_2 sample. SEM analysis demonstrated that both materials present the same microstructure, composed by agglomerated particles of few μm of diameter, which are in turn formed by nanoparticles. Nitrogen adsorption revealed the presence of mesoporosity in both materials, although C,N-TiO_2/SiO_2 presents the highest surface area due to the presence of SiO_2. The presence of nitrogen and carbon in both samples increases their ability to absorb visible light.

Photolysis of PE MPs (not showed here) did not promote changes in the MPs concentration, indicating that the experimental conditions do not promote degradation [9]. Figure 17.2 presents the photocatalytic tests of PE MPs using powdered C,N-TiO_2 at different temperature and pH values [9]. Most of the reaction conditions do not promote degradation of PE MPs, even if the photocatalyst is present in the reaction system. Only at pH 3 and 0 °C, the mass loss achieved 71.77 ± 1.88% [9]. This behaviour was related to both temperature and pH. Cold temperature promotes the fragmentation of the original PE MPs (700–1000 μm) into smaller particles.

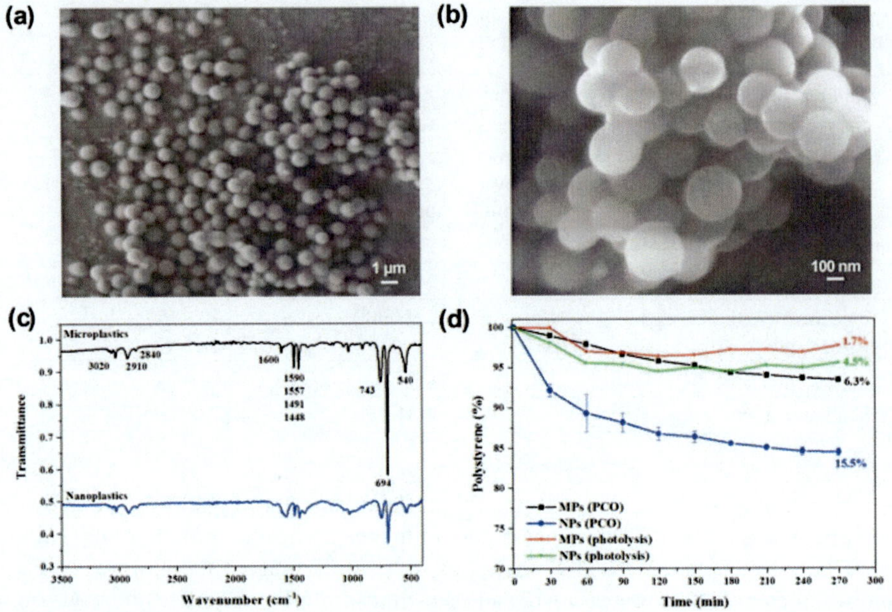

Fig. 17.1 SEM micrographs of the PS **a** MPs and **b** NPs. **c** FTIR spectra of the PS MPs and NPs. **d** Photolysis and photocatalysis of the PS MPs and NPs

This size reduction allowed a better interaction between the plastic and the C,N-TiO_2 powders [9]. On the other side, pH increases the concentration of H^+ ions in the reaction media, favouring the photodegradation mechanism generally proposed for PE plastic [9].

Figure 17.3 shows the degradation of secondary PET MPs by powdered C,N-TiO_2/SiO_2. MPs of $\leq$ 500 μm were used to avoid the use of unrealistic extremely low temperatures (useful for MPs fragmentation) for photocatalytic tertiary treatments at WWTPs. To increase the extension of degradation of the plastic, the photocatalyst with the best characteristic was employed. It was found that secondary PET MPs are a very hard-to-degrade plastic. Indeed, after 120 h of photocatalytic reaction at mild-acidic and mild-alkaline reaction conditions, the highest mass loss achieved was 16%. This result can be attributed to the antioxidants additives that may be contained in original PET plastic.

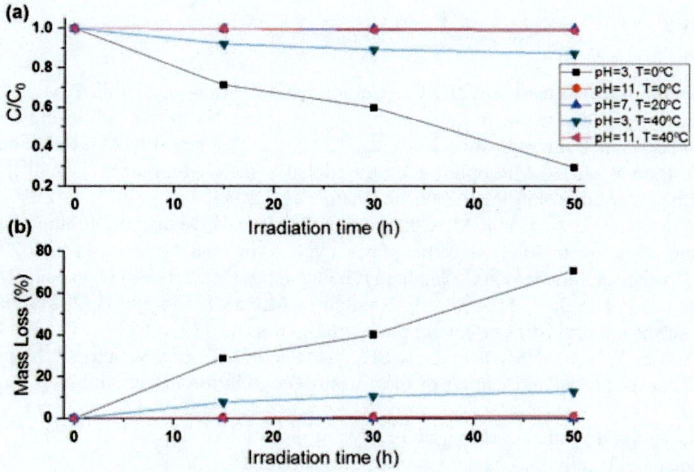

Fig. 17.2 Photocatalysis of PE MPs at different experimental conditions. Reprinted from Journal of Hazardous Materials, 395, M.C. Ariza-Tarazona, J.F. Villarreal-Chiu, J.M. Hernández-López, J. Rivera De la Rosa, V. Barbieri, C. Siligardi, E.I. Cedillo-González, Microplastic pollution reduction by a carbon and nitrogen-doped TiO_2: Effect of pH and temperature in the photocatalytic degradation process, 122632, Copyright (2020), with permission from Elsevier

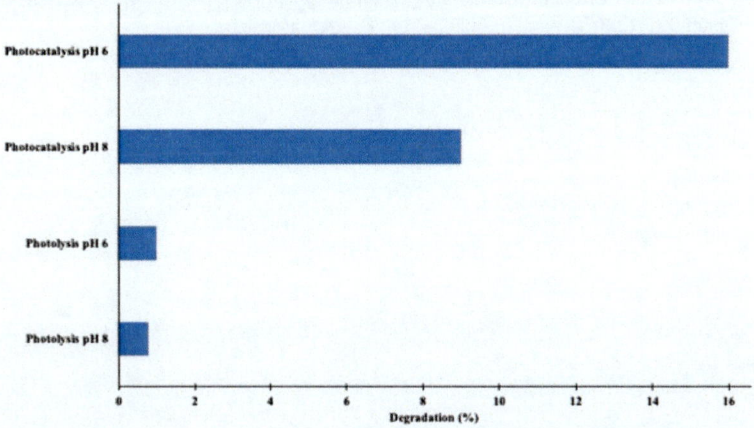

Fig. 17.3 Photocatalytic degradation of PET MPs by C,N-TiO_2/SiO_2 photocatalyst

17.4 Conclusions

MPs and NPs are complex aquatic pollutants that can be degraded by photocatalysis. The properties of the photocatalyst, the MPs and the photocatalytic reaction conditions should be carefully designed in order to achieve degradation efficiency that allow the application of this technology to reduce marine MPs pollution.

References

1. Committee for Risk Assessment (RAC), Committee for Socio-economic Analysis (SEAC) (2000) Background document to the opinion on the annex XV report proposing restrictions on intentionally added microplastics. European Chemicals Agency (ECHA), Helsinki, Finland
2. Barceló D, Picó Y (2019) Microplastics in the global aquatic environment: Analysis, effects, remediation and policy solutions. J Environ Chem Eng 7:103421
3. Brahney J, Mahowald N, Prank M, Cornwell G, Klimont Z, Matsui H, Prather KA (2021) Constraining the atmospheric limb of the plastic cycle. Proc Natl Acad Sci 118:e2020719118
4. Ragusa A, Svelato A, Santacroce C, Catalano P, Notarstefano V, Carnevali O, Papa F, Rongioletti MCA, Baiocco F, Draghi S, D'Amore E, Rinaldo D, Matta M, Giorgini E (2021) Plasticenta: first evidence of microplastics in human placenta. Environ Int 146:106274
5. Leslie HA, van Velzen MJM, Brandsma SH, Vethaak AD, Garcia-Vallejo JJ, Lamoree MH (2022) Discovery and quantification of plastic particle pollution in human blood. Environ Int 163:107199
6. Bioprocess-inspired synthesis of hierarchically porous nitrogen-doped TiO_2 with high visible-light photocatalytic activity. J Mater Chem A 3:19588 (2015)
7. Telford AM, Pham BTT, Neto C, Hawkett BS (2013) Micron-sized polystyrene particles by surfactant-free emulsion polymerization in air: synthesis and mechanism. J Polym Sci, Part A: Polym Chem 51:3997–4002
8. Napper IE, Bakir A, Rowland SJ, Thompson RC (2015) Characterisation, quantity and sorptive properties of microplastics extracted from cosmetics. Mar Pollut Bull 99:178–185
9. Ariza-Tarazona MC, Villarreal-Chiu JF, Hernández-López JM, Rivera De la Rosa J, Barbieri V, Siligardi C, Cedillo-González EI (2020) Microplastic pollution reduction by a carbon and nitrogen-doped TiO_2: Effect of pH and temperature in the photocatalytic degradation process. J Hazard Mater 395:122632

Chapter 18
Determination of Optimal Conditions for Biodegradation of Polystyrene by Bacteria *Bacillus Cereus* and *Pseudomonas Alcaligenes*

Martina Miloloža, Marika Puškarić, Šime Ukić, and Dajana Kučić Grgić

18.1 Introduction

Plastics are extremely versatile due to their excellent properties, which contribute to their widespread production and consumption [1]. The most commonly used for packaging are polypropylene, polyethylene, poly(vinyl chloride), polyurethane, polyethylene terephthalate, and polystyrene (PS). In 2015, 6,300 million tonnes of plastic waste was produced, of which 9% was recycled, 12% was incinerated, and 79% was disposed of in landfills or the environment [1]. However, due to their ubiquitous distribution in marine and freshwater ecosystems, plastics have gradually become a global environmental threat [2]. Natural environmental conditions within these ecosystems, particularly ocean current dynamics, solar radiation, abrasion, and interactions with ships and organisms, cause plastic objects to slowly degrade and fragment into smaller particles called microplastics (MP) [2]. This term is used to describe plastic particles whose longest dimension is less than 5 mm [1, 2]. Adverse effects of MP after ingestion have been observed in aquatic organisms only under laboratory conditions, and usually at very high exposure concentrations that exceed current environmental concentrations by several orders of magnitude [3]. Because of its potentially harmful effects, MP must be removed from the environment [1]. The combination of physical, chemical and biological processes has proven to be the most successful method for removal of MP. Biodegradation is a natural process in which organic matter is used by microorganisms (bacteria, molds, and/or yeasts) as a substrate source, being converted into smaller compounds, water, CO_2, and mineral salts [1]. Nowadays, experimental designs based on response surface methodology

M. Miloloža (✉) · M. Puškarić · Š. Ukić · D. K. Grgić
Faculty of Chemical Engineering and Technology, University of Zagreb, Zagreb, Croatia
e-mail: miloloza@fkit.hr

M. Cocca et al. (eds.), *Proceedings of the 3rd International Conference on Microplastic Pollution in the Mediterranean Sea*, Springer Water,
https://doi.org/10.1007/978-3-031-34455-8_19

are used with purpose to determine the interactions between the independent factors and the dependent variable, and finally to determinate the optimal process conditions [4].

In this work, preliminary experiments were conducted according to the Taguchi design with the aim of determining significant factors, and the main experiment according to the full factorial design to determine the optimal conditions for biodegradation of PS by bacteria *Bacillus cereus* and *Pseudomonas alcaligenes*.

18.2 Experimental

18.2.1 Materials

18.2.1.1 Microplastics

PS particles, in the size range 100–300, 300–500 and 500–700 μm, were obtained by grinding everyday polymer product, such as plastic spoons.

18.2.1.2 Bacteria

Bacillus cereus and *Pseudomonas alcaligenes* were isolated from activated sludge (municipal wastewater treatment Vrgorac, Croatia) and sediment (river Kupa, Croatia), respectively, and used for the biodegradation experiments. These bacteria are stored at Department of Industrial Ecology of Faculty of Chemical Engineering and Technology, University of Zagreb, Croatia.

18.2.2 Methods

18.2.2.1 *Microplastics Preparation*

MP was obtained by grinding plastic accessories in cryo-mill (Retsch, Haan, Germany) accompanied with liquid nitrogen and then dried on the air during 24/48 h at room temperature. Obtained particles were sieved on stainless steel screens (W. S. Tyler RX-86-1 Sieve shaker, USA) to isolate particles in above mentioned size ranges. Before the experiments, MP particles were sterilized in 100-mL flasks containing 70% ethanol during 10 min on rotary shaker (Heidolph unimax 1010, Heidolph incubator 1000, Germany) at 160 rpm. Particles were separated from the ethanol by vacuum membrane filtration and additionally washed with sterile deionised water.

18.2.2.2 Bacterial Cultivation

Bacteria were cultivated on nutrient agar at 37 °C during 24 h. Cultures were pre-cultured in mineral media for 24 h prior to the experiment to adapt to conditions with mineral media [5]. Optical densities (OD) of bacterial suspensions were measured at $\lambda = 600$ nm by spectrophotometer (Hach, USA) and CFU was monitored by decimal plate method [6].

18.2.2.3 Biodegradation Experiments

Two sets of biodegradation experiments were performed: the preliminary experiment designed by Taguchi and the main experiment designed by full factorial plan.

Biodegradation of PS particles were investigated by bacteria *B. cereus* and *P. alcaligenes*. Experiments were performed in Erlenmeyer flasks with a working volume of 100 mL for 30 days in a rotary shaker. The flasks contained mineral media [5], bacterial suspension and MP-PS. Control flasks without MP-PS were set in the purpose of monitoring bacterial growth.

According to Taguchi design of experiments, 7 factors at 2 levels (Table 18.1) were investigated. Experiments were marked as P1-1-8 and P2-1-8 for *B. cereus* and *P. alcaligenes*, respectively.

Table 18.2 shows a combination of the 3 factors at 3 levels (full factorial design). Constant parameters were: pH $= 7.0 \pm 0.3$, $T = 25.0 \pm 0.1$ °C, without glucose addition (γ_{GLU}), except that OD $= 0.3$ and $\gamma_{MP} = 500$ mg/L for experiment with *B. cereus* and *P. alcaligenes*, respectively. The experiments were labelled as P1-1-27 and P2-1-27 for MP-PS biodegradation by *B. cereus* and *P. alcaligenes*, respectively.

Analysis of the most important process parameters as well as optimal process conditions for biodegradation of PS by *B. cereus* and *P. alcaligenes* were determined using CFU as response parameter. The most statistically significant factors and optimal process conditions were determined by analysis of variance effects

Table 18.1 Taguchi L_8 Orthogonal Array for PS biodegradation by *B. cereus* and *P. alcaligenes*

Experiment	pH-value (–)	T (°C)	Size of MP (μm)	γ_{MP} (mg/L)	Agitation speed (rpm)	OD (–)	γ_{GLU} (mg/L)
P1-1; P2-1	8	15	500–700	1000	100	0.5	0
P1-2; P2-2	6	15	100–300	50	100	0.1	0
P1-3; P2-3	8	15	100–300	50	200	0.1	100
P1-4; P2-4	6	15	100–300	1000	200	0.5	100
P1-5; P2-5	6	25	500–700	50	100	0.1	100
P1-6; P2-6	8	25	100–300	1000	100	0.1	100
P1-7; P2-7	6	25	500–700	1000	200	0.1	0
P1-8; P2-8	8	25	100–300	50	200	0.5	0

Table 18.2 Full factorial design for PS biodegradation by *B. cereus* and *P. alcaligenes*

	Bacillus cereus			*Pseudomonas alcaligenes*		
Experiment/ Factor	γ_{MP} (mg/L)	Size of MP (μm)	Agitation speed (rpm)	OD (-)	Size of MP (μm)	Agitation speed (rpm)
P1-1; P2-1	1000	300–500	100	0.1	300–500	200
P1-2; P2-2	50	300–500	200	0.5	300–500	100
P1-3; P2-3	1000	100–300	150	0.3	100–300	200
P1-4; P2-4	50	100–300	100	0.1	100–300	100
P1-5; P2-5	500	300–500	100	0.1	300–500	150
P1-6; P2-6	50	500–700	150	0.3	500–700	100
P1-7; P2-7	50	500–700	100	0.1	500–700	100
P1-8; P2-8	500	100–300	100	0.1	100–300	150
P1-9; P2-9	500	500–700	100	0.1	500–700	150
P1-10; P2-10	1000	100–300	200	0.5	100–300	200
P1-11; P2-11	1000	100–300	100	0.1	100–300	200
P1-12; P2-12	50	100–300	150	0.3	100–300	100
P1-13; P2-13	500	300–500	200	0.5	300–500	150
P1-14; P2-14	500	300–500	150	0.3	300–500	150
P1-15; P2-15	1000	500–700	150	0.3	500–700	200
P1-16; P2-16	50	500–700	200	0.5	500–700	100
P1-17; P2-17	50	100–300	200	0.5	100–300	100
P1-18; P2-18	1000	500–700	200	0.5	500–700	200
P1-19; P2-19	1000	500–700	100	0.1	500–700	200
P1-20; P2-20	500	500–700	150	0.3	500–700	150
P1-21; P2-21	500	500–700	200	0.5	500–700	150
P1-22; P2-22	1000	300–500	200	0.5	300–500	200
P1-23; P2-23	1000	300–500	150	0.3	300–500	200

(continued)

Table 18.2 (continued)

	Bacillus cereus			*Pseudomonas alcaligenes*		
Experiment/Factor	γ_{MP} (mg/L)	Size of MP (μm)	Agitation speed (rpm)	OD (-)	Size of MP (μm)	Agitation speed (rpm)
P1-24; P2-24	50	300–500	100	0.1	300–500	100
P1-25; P2-25	50	300–500	150	0.3	300–500	100
P1-26; P2-26	500	100–300	150	0.3	100–300	150
P1-27; P2-27	500	100–300	200	0.5	100–300	150

(ANOVA) using Design-Expert 10.0, StatEase, USA. Statistical significance of the factors effects was considered at a 95% confidence level.

18.2.2.4 Analytical Techniques

During the experiments, CFU, total carbon (TC), total organic carbon (TOC), inorganic carbon (IC) (TOC-VCSH, Shimadzu) were determinate. At the end of the experiments, MP-PS was separated from the aqueous phase by membrane filtration and particles were analysed by FTIR-ATR (Spectrum One, PerkinElmer) spectroscopy in order to confirm the biodegradation.

18.3 Results and Discussion

18.3.1 Results

The main purpose of the preliminary experiment was to determine the key factors (listed in Table 18.3) affecting the biodegradation process of PS particles by *B. cereus* and *P. alcaligenes*.

According to the key factors, the optimal conditions, shown in Table 18.4, for biodegradation of PS were investigated by full factorial design.

During the main experiment, changes of CFU, TOC and IC values were monitored as it is shown in Fig. 18.1 and Table 18.5.

At the end of the experiment, PS particles were analysed by FTIR-ATR spectroscopy, which is shown in Fig. 18.2.

Table 18.3 Significant factors for PS biodegradation by *B. cereus* and *P. alcaligenes* obtained by Taguchi design

Factor	Sum of squares	DF	*F* value	*p*-value	Contribution (%)
	Bacillus cereus				
γ_{MP}	0.07	1	128.82	0.002	5.03
Size of MP	0.68	1	1243.26	0.000	48.52
Agitation speed	0.58	1	1058.59	0.000	41.31
	Pseudomonas alcaligenes				
OD	0.49	1	12.22	0.040	13.91
Size of MP	0.51	1	12.72	0.038	14.48
Agitation speed	1.88	1	46.93	0.006	53.42

Table 18.4 Optimal conditions for PS biodegradation by *B. cereus* and *P. alcaligenes* obtained by full factorial design

	Bacillus cereus			*Pseudomonas alcaligenes*		
Factors	γ_{MP} (mg/L)	size of MP (μm)	agitation speed (rpm)	OD (–)	size of MP (μm)	agitation speed (rpm)
Value	66.20	413.29	100.45	0.35	334.73	161.08

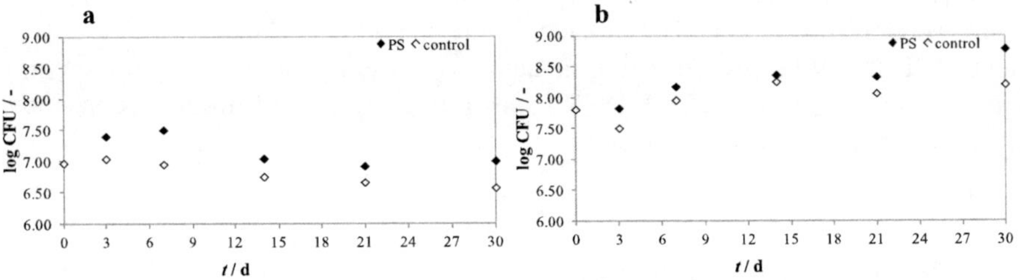

Fig. 18.1 Changes of log CFU of *B. cereus* **a** in P1-24 and *P. alcaligenes* **b** in P2-23 during 30 days of biodegradation of PS

Table 18.5 Changes of TOC and IC for blank probe (BP) and sample (PS) during 30 days of biodegradation of PS by *B. cereus* (P1-24) and *P. alcaligenes* (P2-23) by full factorial design

	Bacillus cereus				*Pseudomonas alcaligenes*			
t/d	TOC (BP)/(mg/L)	TOC (PS)/(mg/L)	IC (BP)/(mg/L)	IC (PS)/(mg/L)	TOC (BP)/(mg/L)	TOC (PS)/(mg/L)	IC (BP)/(mg/L)	IC (PS)/(mg/L)
0	2.51	2.51	0.76	0.76	0.62	0.62	0.22	0.22
3	2.10	2.42	0.10	0.87	4.31	5.52	0.31	1.04
7	4.12	4.99	0.30	0.88	6.84	7.74	0.95	2.91
14	3.90	4.22	0.19	0.37	8.02	10.03	2.11	4.71
21	3.37	4.67	0.23	0.44	6.71	7.80	1.25	2.07
30	3.61	4.64	0.25	0.40	6.98	8.78	1.48	3.23

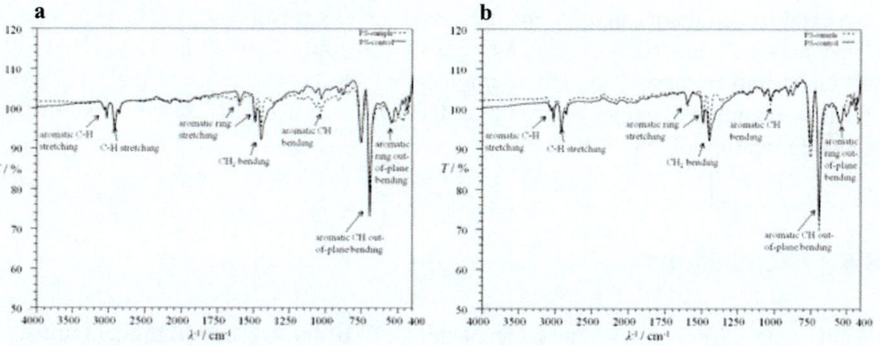

Fig. 18.2 FTIR-ATR spectroscopy of PS particles before (PS control) and after (PS sample) biodegradation by *B. cereus* **a** in P1-24 and *P. alcaligenes* **b** in P2-23

18.3.2 Discussion

Significant factors for biodegradation of PS by *B. cereus* were γ_{MP}, size of MP, and agitation speed (Table 18.3); the largest contribution was shown for the size of PS particles. This is not surprising as smaller size and higher surface area contribute to bacterial colonization on MP particles [7]. The agitation speed of the rotary shaker maintains the concentration of dissolved oxygen, which is essential for biochemical reactions [8]. According to the oxygen demand, *B. cereus* is a facultative anaerobe, which means that it can grow under anoxic conditions [9]. In addition, this rod-shaped bacterium forms endospores that make it more resistant to extreme environmental conditions [9]. The above factors were also important for PS biodegradation by Gram negative *P. alcaligenes*, but the third factor was OD. The higher OD indicates higher CFU, which consequently increases the efficiency of biodegradation. According to the mentioned significant factors, the main experiment was conducted and optimal conditions for PS biodegradation by *B. cereus* can be achieved at low PS concentration, average PS particle size and at 100 rpm (Table 18.4). On the other hand, the average PS particle size at higher stirring speed (*P. alcaligenes* is an obligate aerobe, which means it grows in an oxygen-rich environment [10]) and an average value of OD were the most suitable conditions for *P. alcaligenes* growth (Table 18.4). During the main experiment, CFU and TOC values were monitored in the purpose of better understanding of the process. Namely, changes of CFU and TOC had the same trend; values increased during 7 and 14 day in samples with *B. cerues* and *P. alcaligenes*, respectively (Fig. 18.1 and Table 18.5). In the blank probe, TOC values were also elevated, indicating bacterial cells lysis [11]. However, an increase in TOC in the sample (PS) may indicate the release of additives from the surface of particles. The increase in TOC in the sample correlates with an increase in IC values, indicating that biodegradation of PS has occurred [11]. According to the higher values of CFU and TOC, *P. alcaligenes* had the higher potential of PS biodegradation. At the end of the experiment, the particles were analysed by FTIR spectroscopy. The characteristic peaks for PS are 3024 cm^{-1}, 2847 cm^{-1}, 1601 cm^{-1}, 1492 cm^{-1}, 1451 cm^{-1}, 1027 cm^{-1} and 694 cm^{-1} [12]. The intensity of the peaks for the sample PS was much lower compared to the control PS, as it can be seen in Fig. 18.2, which confirms the biodegradation of PS.

18.4 Conclusions

Overall, agitation speed, size and concentration of MP and OD had significant impact on biodegradation process. According to this research, *Pseudomonas alcaligenes* showed higher capability for PS biodegradation in comparison to *B. cereus.*

Acknowledgements The authors would like to acknowledge the financial support of the Croatian Science Foundation through project entitled Advanced Water Treatment Technologies for Microplastics Removal (AdWaTMiR) (IP-2019-04-9661).

References

1. Miloloža M (2022) Biotreatment strategies for the removal of microplastics from freshwater systems. A review. Environ Chem Lett 20:1377–1402
2. Li J (2018) Microplastics in freshwater systems: a review on occurrence, environmental effects, and methods for microplastics detection. Water Res 137:362–374
3. Issac MN (2021) Effect of microplastics in water and aquatic systems. Environ Sci Pollut Res 28:19544–19562
4. Morgan E (1989) Practical exploratory experimental designs. Chemom Intell Lab Syst 5:283–301
5. Kyaw BM (2012) Biodegradation of low density polyethene (LDPE) by *Pseudomonas* Species. Ind J Microbiol 52:411–419
6. Black JG (2012) Microbiology—principles and explorations. Wiley, Hoboken
7. Ali SS (2021) Plastic wastes biodegradation: mechanisms, challenges and future prospects. Sci Total Environ 780:146590
8. Schmidt-Rohr K (2020) Oxygen is the high-energy molecule powering complex multicellular life: fundamental corrections to traditional bioenergetics. ACS Omega 5:2221–2233
9. El-Arabi TF (2013) Characteristics of *Bacillus cereus*. In: JG Morris, E Morris (eds) Foodborne infections and intoxications. Elsevier, Amsterdam, pp 401–407
10. Suzuki M (2013) Genome sequence of a strain of the human pathogenic bacterium *Pseudomonas alcaligenes* that caused bloodstream infection. Genome Announc 1:e00919-e1013
11. Kučić Grgić D (2021) Bioremediation of MP-polluted waters using bacteria *Bacillus licheniformis, Lysinibacillus massiliensis*, and mixed culture of *Bacillus* sp. and *Delftia acidovorans*. Chem Biochem Eng Q 35:205–224
12. Jung MR (2008) Validation of ATR FT-IR to identify polymers of plastic marine debris, including those ingested by marine organisms. Mar Pollut Bull 127:704–716

Chapter 19
Microplastics in Drinking Water: Italian National Working Group and Analytical Methods

Lorenzo Martellone, Eleonora Brancaleone, Daniela Mattei, Valentina Fuscoletti, Luca Lucentini, Gabriele Favero, Gerardo Califano, Anna Luisa Salvatore, Tania Tellini, Giuseppe Bortone, and Gruppo Nazionale di Lavoro Microplastiche

19.1 Introduction

Microplastics are "emerging" heterogeneous contaminants with a complex toxicological profile. Extremely widespread in different water bodies, microplastics have been studied for some time due to the possible effects on human and animal health. Drinking waters could be a potential source of microplastics due to the possible surface and ground waters contamination [1]. According to Directive (EU) 2020/2184 [2] on the quality of water intended for human consumption, the European

L. Martellone (✉) · E. Brancaleone · D. Mattei · V. Fuscoletti · L. Lucentini
Department of Environment and Health, Italian National Institute of Health (ISS), Rome, Italy
e-mail: lorenzo.martellone@uniroma1.it

L. Martellone
Department of Chemistry & Technology of Drugs, Sapienza University of Rome, Rome, Italy

E. Brancaleone · G. Favero
Department of Environmental Biology, Sapienza University of Rome, Rome, Italy

G. Califano · A. L. Salvatore
Office 4—Chemical, Physical and Biological Risk Prevention and Environmental Health Promotion, Health and Safety Protection, Ministry of Health, Rome, Italy

T. Tellini
Water Sector, Utilitalia, Rome, Italy

G. Bortone
ASSOARPA, Rome, Italy

G. Bortone · Gruppo Nazionale di Lavoro Microplastiche
ARPAE, Bologna, Italy

M. Cocca et al. (eds.), *Proceedings of the 3rd International Conference on Microplastic Pollution in the Mediterranean Sea*, Springer Water,
https://doi.org/10.1007/978-3-031-34455-8_20

Commission will adopt a methodology to measure microplastics with the aim of including them on the watch list, in order to respond at the growing concerns related to drinking water in a dynamic and flexible way. On this basis, the Joint Research Centre (JRC) launched a scientific project finalized to harmonize experience and knowledge about microplastic analysis in drinking water, requiring support from the national technical-scientific representatives. So, the Italian National Institute of Health (ISS), with the coordination of the Italian Ministry of Health, defined a National Working Group which includes experts from the National Research Council (CNR), from the national network of local environmental Authorities (the National System for Environmental Protection, SNPA, consisting of the Italian Institute for environmental protection and research ISPRA and the various ARPA, local environmental protection agencies), Universities and Water Suppliers (Utilitalia). This group was designed to work on: (i) JRC and EC support on national expertise about microplastic monitoring in drinking water (ii) development of a national analytical method for microplastics in drinking water to be presented to JRC.

The recruitment phase took place on March 2022, with a "call to action" [3] by the Italian National Institute of Health and following discussions with interested stakeholders. The inclusion criteria were: (i) experience in microplastic analysis regardless the matrix type (SNPA, Universities, CNR), (ii) proven expertise on emerging contaminants and involvement in the issue (Water Suppliers). Therefore, after collecting applications and checking inclusion criteria, with the agreement of the Ministry of Health, the group was formed. The group, apart from representatives of ISS and Ministry of Health, consist of 41 members: 18 from SNPA: ISPRA and ARPA, 12 from Water Suppliers and 11 from Universities and CNR. Once the group was assembled, the first meeting was organized in June 2022.

19.2 Proceedings of the First Meeting

The first meeting of Italian National Working Group for microplastics in Drinking Water took place on June 16th, 2022.

During the first meeting, participants were invited to introduce themselves giving an overview on the overall experience in microplastic analysis, including sampling techniques, pre-treatment methods and instrumental analysis. The information requested from group members were based on a JRC survey, published online in early 2022. As some members of the working group collaborate, they present themselves as a working team (23 in total).

Representatives of ISS and Ministry of Health explained the purposes of the project and advances in the Italian transposition of Directive (EU) 2020/2184 on the quality of water intended for human consumption. SNPA coordinator explained issues in microplastics analysis that various members of ARPA had in the past years while Utilitalia coordinator illustrated the state of knowledge and expertise on the topic by water suppliers.

ISS focused on the new mechanism of "watch list" for emerging contaminants and the problems related to microplastic definition.

The "watch list", described as a complex but useful tool, includes substances and compounds that are likely to be present in water intended for human consumption and could pose a potential risk to human health; the "watch list" is periodically updated with the inclusion of new compounds. Microplastics will be added in the "watch list" if inclusion criteria (analytical method development and risk analysis results) are met. To that end, the European Commission shall make use, in particular, of scientific research of the World Health Organization (WHO).

The problem of microplastic definition, also explained by ISS in this section of the meeting, emerged from the particularity of these contaminants and was reaffirmed citing the opinions of European Chemicals Agency (ECHA) committees in the context of a proposed restriction of microplastics intentionally added in a wide range of products [4, 5]. Have also been mentioned, as being directly related to the problem of plastic pollution, Directive (EU) 2019/904 ("Single Use Plastic" Directive) [6] and the Italian legislative act for microplastics "microbeads" ban in cosmetics [7]. The Utilitalia representative confirmed the complex nature of microplastic analysis but also reported that water suppliers, thanks to pre-existing internal working groups, have been working on a common analytical method for several months. Utilitalia representative also underlined the need for a routine analytical method, i.e., a method that is scientifically valid but also easily suitable for water suppliers during drinking water quality analysis.

The conference ended with a panel discussion regarding scientific priorities, common objectives and planning of group activities in the coming months.

19.3 Results and Discussion

This first meeting of the Italian National Working Group showed how, despite there is a diversified knowledge among the different institutions, the magnitude of the problem has been well understood by all the members, even if some of them, especially water suppliers, have approached it only recently. From the contributions presented by Italian National Working Group members, it can be observed that there are heterogeneous analytical strategies but with common elements; experts, despite belonging to different organizations with different roles and objectives, are extensively working on the water matrix (Fig. 19.1), particularly on surface water, drinking water and sea water. Issues of common interest are also wastewater, for the possibility of massive contamination from civic and industrial sources and groundwater, as hypothetically less contaminated from microplastics.

Moreover experts, especially groups belonging to public institutions, are expanding their research lines also to other types of matrices such as biota or sediments.

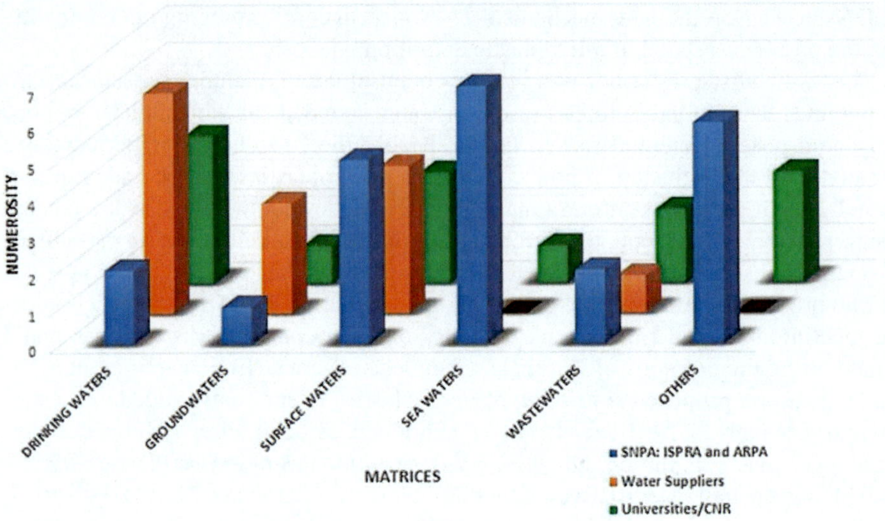

Fig. 19.1 Distribution of studies related to the presence of microplastics in the different matrices analyzed among the various authorities and institutions belonging to the working group

Drinking water is obviously the most frequently studied matrix by water suppliers, but many of them are extending their research to surface water and groundwater, as waters intended for drinking water production and possible source of microplastics.

On the contrary, national, and local environmental Authorities focused more on sea water as part of monitoring related to marine strategy [8]; ARPA and ISPRA were also involved in sediment and biota analysis, surface water and groundwater. Only a small percentage of ARPA and ISPRA researchers were studying drinking water when they were recruited, even though interest in the development of an analytical method for drinking water was rapidly growing among them.

Universities and CNR are mostly involved than ARPA and ISPRA in drinking water analysis, with also a focus on surface water, other matrices (sediments, biota) and wastewater. Only a small percentage focused on sea water and groundwater.

These choices, as a direct consequence, affects the development of sampling techniques and analytical methods; in fact, clusters of sampling methods can be identified from the survey according to the type of water sampled. Manta net appears to be the most suitable sampling method for seawater samples; all the group members who approached the issue of marine pollution originating from microplastics refers to Manta net as their standard sampling method.

Indeed, for other sample types, the situation appears different and more varied: in these cases, other sampling methods, such as glass bottles or "in situ-filtration methods" i.e., systems connected to sieves/filters, have been developed. These systems allow filtration of large volumes of water without sample loss but significant increases in costs and sampling time [9].

For drinking water, the most common solutions appear to be sampling with bottles for water suppliers while Universities and CNR are also interested in trying alternative solutions as "in situ-filtration methods". This *status* describes the different objectives and roles of the organization involved. Water suppliers need a simple, quick sampling method and bottle sampling from taps, despite low volumes of water collectable, are obviously preferred. On the contrary, researchers from universities and CNR are interested in more complex methods for representativeness (high volumes of water filtered) and lowering-contamination (closed systems) reasons. It should also be stated that, for a decent numbers of working group members, differences between sampling methods are not well understood and/or sampling is not directly performed by them.

The survey trend also shows that working group members are developing analytical protocols for microplastics with or without a pre-treatment phase. An untreated sample, especially full of organic matter samples (surface water, wastewater) could be problematic in terms of signal-to-noise *ratio* and goodness of results but not requires effectiveness and plastic-degradation studies and could globally speed up the entire analytical procedure. On the other hand, with dedicate pre-treatment protocols, results could be significantly better, but operations and the development of the procedure could be also time-consuming. In this respect, it is noted that groups have chosen the method by pre-treating protocols, opted for (i) digestion with, for example, acids or hydrogen peroxide or (ii) digestion followed by an extraction step.

As seen in sampling phase, the water type influences the choice of analytical criteria, so pre-treatment methods depend on what kind of water working group members analysed.

Sea water samples, mainly retrieved by manta net sampling and conducted by ARPA members, are usually dried, sieved and submitted to digestion protocols (hydrogen peroxide, with or without heating). For groups who experienced drinking water analysis, a pre-treatment step with light digestion (acids or peroxides) without temperature programs and followed by a filtration on a substrate is very common. The most popular substrates for drinking water filtration are silicon square filters, which have proven to be optimal for Raman and Fourier Transform Infrared (FTIR) microscopy analysis; alumina round filters have also been used with good results in Raman and FTIR microscopy analysis. Silicon filters have a small surface area that simplifies particle-based analysis in microscopy coupled to vibrational spectroscopy techniques but require a low volume of water to be filtered in order to reduce overlapping of particles and matrix interference.

On the contrary, surface water samples followed the direction of the distinctive research group: analysis of surface water carried out by ARPA members followed the marine strategy protocols (manta trawl—drying—sieving—digestion) while analysis of surface water carried out by water suppliers are usually carried out in a similar way of drinking water (bottle sampling—light digestion—filtration on substrate); only few authors carried out density separation (with or without digestion) using an hyper-saline solution.

Ultimately, even from an instrumental point of view, there is heterogeneity. As reported by the survey, the most frequently used analytical techniques for the

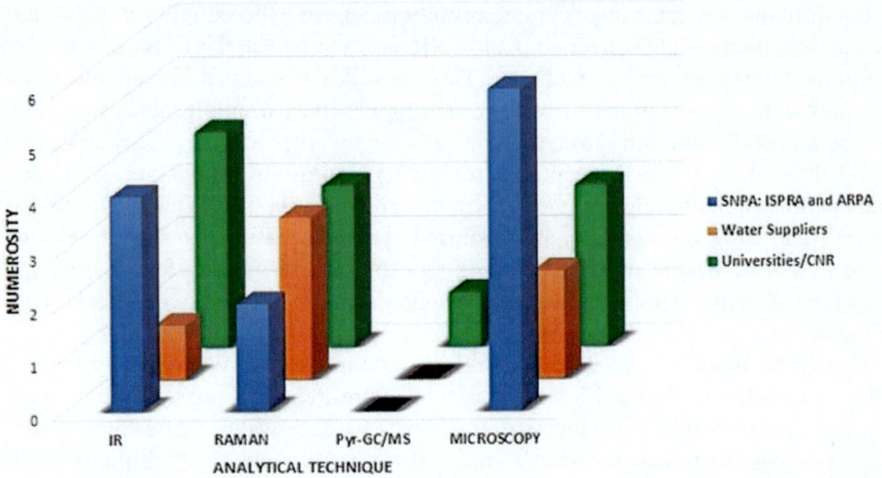

Fig. 19.2 Distribution of selected analyical tecniques in microplastics analysis among the various authorities and institutions belonging to the working group

characterization of microplastics were Microscopy (optical microscopy and electron microscopy), FTIR spectroscopy (normal and microscopy-coupled), Raman microscopy and Gas Chromatography–Mass Spectrometry coupled to Pyrolysis (Py-GC–MS), as shown in Fig. 19.2.

The employment of spectroscopy (FTIR and Raman) was very common among working groups regardless of affiliation. Microscopy-only techniques are also frequently used, especially from ISPRA and ARPA members. On the contrary, Gas Chromatography–Mass Spectrometry coupled to Pyrolysis, as an emerging technique in microplastic analysis, is currently used only by a single group affiliated to university. It is also necessary to consider that about 17% of working teams still do not have the instrumentation for microplastic analysis in drinking water although most of them are moving to purchase and/or are waiting for suggestions from other group members for the purchase. This is explained by the very recent interest of some ISPRA and ARPA members to drinking water analysis and of some water suppliers in the topic.

The survey also confirmed that the choice of analytical techniques follows the type of water investigated. In this context, 3 major clusters can be observed: sea water samples, surface water samples and drinking water samples. Sea water samples are usually observed with microscopy (optical or electron microscopy), as they include large microplastics (>300 μm) retrieved from Manta nets. Some research groups also confirmed the identity of some polymers with FTIR spectroscopy.

Drinking water samples, as they often originate from water filtered on a substrate, are usually analysed with microscopy coupled to spectroscopy techniques as FTIR microscopy and Raman microscopy. Water suppliers usually preferred Raman over FTIR given its ability to be insensitive to water vibrational signals during the analysis.

Surface water samples, as seen in pre-treatment phase, followed the direction of the distinctive research group: ARPA members usually performed microscope analysis while water suppliers were prone to use FTIR microscopy and Raman microscopy.

From the panel discussion and survey suggestions, scientific priorities, and common objectives to be pursued emerged. The two major issues appeared to be variability in results and the relative lack of microplastics certified standard; also blanks preparation and contamination prevention were discussed. Variability, as suggested by various group members, could be minimized by using similar analytical techniques with a standard procedure; also certified microplastic standards could be useful. Some working group researchers in the past experienced proficiency tests with certified standard and agreed to share information about where to find/how to fabricate them. In term of contamination prevention, it was stated that is virtually impossible to prevent every source of contamination and is important to provide environmental blanks, procedural blanks and reagent blanks in order to avoid results overestimation.

19.4 Conclusions

As underlined by the results of the first national survey and working group members' contributions, the *status* of microplastic analysis in Italy appears to be heterogeneous and constantly evolving. Researchers from organizations with different roles and objectives were already working on an analytical method for microplastics identification and quantification in drinking water, but without agreement on sampling method, pre-treatment, or analytical technique. The main difficulty that has emerged is to align the different targets of public research authorities and agencies and water suppliers. To overcome this issue, it will be necessary to develop a method that takes into account the needs of both, and, in particular, that is routinely applicable. Working group, as well as for method development, will be also useful to share know-how and data, waiting for more information about health implication of microplastics. In line with the working group aims, during the next months the participants will contribute to the drafting of a technical paper (ISTISAN Report) focusing on analytical techniques for the analysis of microplastics, while starting to work on the common analytical method. The next scheduled appointments are:

- The participation of ISS, SNPA and Utilitalia representatives, coordinators of the working group, to the three workshops planned by JRC in September 2022. During these workshops will be shared the results of European JRC survey, about the state of knowledge on microplastics;
- The next working group meeting, scheduled for November 2022,that will be focus also on these results.

References

1. Marsden P, Koelmans B, Bourdon-Lacombe J, Gouin T, D'Anglada L, Cunliffe D, Jarvis P, Fawell J, De France J (2019) Microplastic in drinking water. World Health Organization, Geneva
2. Directive (EU) 2020/2184 of the European Parliament and of the Council of 16 December 2020 on the quality of water intended for human consumption (recast). Offic J Europ Union L 435/1, 23 December 2020
3. Microplastiche nelle acque potabili, 'call to action' per i contributi italiani alle azioni europee. Istituto Superiore di Sanità (In Italian). https://www.iss.it/news/-/asset_publisher/gJ3hFqMQsykM/content/id/6776622
4. Proposal for a restriction: intentionally added microplastics (Annex XV restriction report: Version 1.2). Helsinki: European Chemical Agency (ECHA), Helsinki (2019)
5. Opinion of the Committee for Risk Assessment and Opinion of the Committee for Socio-economic Analysis on an Annex XV dossier proposing restrictions on intentionally-added microplastics. European Chemical Agency (ECHA), Helsinki (2020)
6. Directive (EU) 2019/904 of the European Parliament and of the Council of 5 June 2019 on the reduction of the impact of certain plastic products on the environment. Offic J Europ Union L 155/1, 12 June 2019
7. Legge 27 dicembre 2017, n. 205. Bilancio di previsione dello Stato per l'anno finanziario 2018 e bilancio pluriennale per il triennio 2018–2020 (17G00222) (In Italian). Gazzetta Ufficiale - Serie Generale n. 302, 29 dicembre 2017
8. Directive 2008/56/EC of the European Parliament and Of The Council of 17 June 2008 establishing a framework for community action in the field of marine environmental policy (Marine Strategy Framework Directive). Offic J Europ Union L 164/19, 17 June 2008
9. Martellone L, Lucentini L, Mattei D, De Vincenzo M, Favero G, Bogialli S, Litti L, Meneghetti M, Corami F (2021) Rosso, B: Strategie di campionamento di microplastiche negli ambienti acquatici e metodi di pretrattamento (Rapporti ISTISAN 21/2). Istituto Superiore di Sanità, Rome

Chapter 20
Ecotoxicological Assessment of Alento River Catchment a Focus on Microplastic Impact

Sara Accardo, Simona Schiavo, Luisa Parrella, Angelo Raffaele Di Maiolo, Marina Paolucci, Mariano Mercurio, Cristiano Gramegna, Vincenzo Pasquale, and Sonia Manzo

20.1 Introduction

Human activities often generate contaminants that can compromise the function of ecosystems. Rivers in particular collected and often transported several kinds of chemicals and items producing a mixture that can generate adverse impacts on the biota. Dams are often present to regulate water for irrigation, hydropower generation, flood control, and drinking purposes. Besides, they are known to trap pollutants that accumulate within their reservoirs. More attention is paid to microplastics (MPs, i.e., plastic particles < 5 mm), an emerging contaminant in waterways worldwide, and how they move along rivers, whether MPs also accumulate behind dams is an important question for informing estimates of global river inputs to coastal areas [1]. Actually, MPs may also settle in the sediments [2] in the function of their density, and physical, chemical, and biological factors [3]. They represent potential physical and chemical threats since their impact is linked not only to mechanical injuries but also to the leaching of chemicals.

S. Accardo (✉) · S. Schiavo · L. Parrella · A. R. Di Maiolo · S. Manzo
ENEA CR PORTICI, P. Le E. Fermi 1, 80055 Portici, NA, Italy
e-mail: sara.accardo001@studenti.uniparthenope.it

S. Accardo · V. Pasquale
Department of Science and Technology, Parthenope University of Naples, Naples, Italy

M. Paolucci · M. Mercurio
Department of Sciences and Technologies, University of Sannio, Via Port'Arsa, 11, 82100 Benevento, Italy

C. Gramegna
ARPAC—Via Vicinale Santa Maria del Pianto, Centro Polif. TORRE, 80143 Napoli, Italy

M. Cocca et al. (eds.), *Proceedings of the 3rd International Conference on Microplastic Pollution in the Mediterranean Sea*, Springer Water,
https://doi.org/10.1007/978-3-031-34455-8_21

The ecotoxicological assessment of environmental samples allows the evaluation of the integrated effects of all present contaminants, evidencing joint action of pollutants (additive, synergistic, and antagonistic). It is generally based on a battery of bioassays with organisms belonging to several species since a species sensitive to all environmental contaminants does not universally exist [4, 5].

The Alento River catchment (Southern Italy) is located between the Cilento, Vallo di Diano and Alburni National Park (Campania, southern Apennine). It represents an interesting case study partly because of the presence of an integrated system of storage dams devoted to agriculture and tourism and due to the presence of high biodiversity, covered mainly by shrubs and forests.

In this contribution, a preliminary evaluation of the ecotoxicological status of the Alento River, with a special focus on MPs occurrence and associated ecological risk, is presented.

20.2 Materials and Methods

20.2.1 Study Area and Sample Collection

The Alento River catchment is situated in the Campania region, Southern Italy, and lies within the Cilento area, with a total drainage area of about 411 km^2 and a perimeter of approximately 145 km. A sampling campaign of water (FW) and sediment (S) was conducted in 2020, at five different sites along the river (upstream to downstream: Al1, OAl1, Al3, Al4bis, and Al5) (Fig. 20.1). Besides, at Al4bis also fishes were sampled and analyzed for their content in MPs.

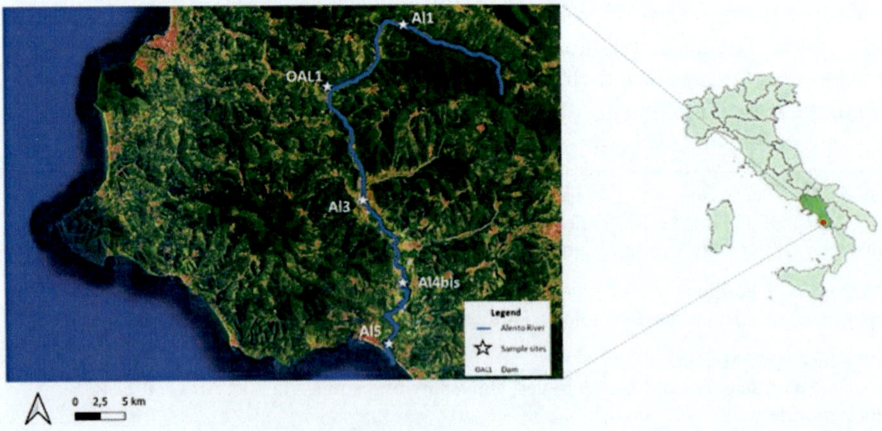

Fig. 20.1 Study area and sampling sites of Alento river

20.2.2 Ecotoxicological Bioassays

Freshwater samples and sediment elutriates (ISPRA, MLG 67/2011) [6] were tested using a battery of toxicity tests with model organisms belonging to the three main trophic levels: *Vibrio fischeri* (bioluminescence inhibition test ISO 11348–1997) [7], *Selenastrum capricornutum* (growth inhibition test, OECD 201 2006) [8], *Daphnia magna* (immobilization test, OECD 202 2004) [9] and *Thamnocephalus platyurus* (acute immobilization test, ISO 14380:2011) [10].

Data were expressed as effect percentage with respect to the control by using the Abbott formula. Results were integrated to obtain a Toxicity Battery Index (TBI) as described in Manzo et al. (2014) [11].

20.2.3 Microplastic Extraction and Identification

Water samples (1L) were filtered using a vacuum pump on Whatman GF/C (Glass microfiber filter 1.2 μm) filter paper [12].

For sediment samples, the density separation method to extract microplastics was used [13]. The collected wet sediment samples were oven-dried 40 °C for 24 h and pre-treated with several variable mesh sizes sieves to reduce the sample volume. 30 g (dw) was added to 200 mL of a saturated $NaCl_2$ solution and stirred for 20 min and then left to settle. After 24 h, the supernatant was filtered and filter papers were washed several times with deionized water. MPs were counted and categorized according to color and morphology by Zeiss stereomicroscope. In the case of fish samples, the MPs extracted from the gut were analyzed using Bruker's FTIR ALPHA spectrometer. The acquired spectra were processed using Bruker OPUS ver software 7.2.

The abundance of MPs is expressed as particles/L for waters and in particles /g for sediments. GIS (13.4, Esri, USA) was used to describe the geographic location.

20.2.4 Ecological Risk Assessment

Probabilistic ERA has been accomplished according to the method proposed by Solomon et al. (1996) [14] and Hall et al. (1999) [15], comparing exposure concentration distribution with the 5th percentile of the corresponding effect distribution.

A collection of toxicity data from open literature sources, and an international electronic database, were performed to define the MPs sensitivity distribution of the organisms.

The different dose descriptors (i.e., LC10-20–50, EC10-20–50, LOEC) were converted into chronic NOECs [16]. The 5th percentile, representing the effect threshold limit ensuring the protection of 95% of the species from MPs adverse effects, was estimated from the regression equation. The results were expressed as the probability that MPs exposure concentration in water and sediment will exceed the corresponding 5th percentile.

20.3 Results and Discussion

20.3.1 Ecotoxicological Assays

The results related to the ecotoxicological evaluation of water and sediment samples are reported in Fig. 20.2. Regarding water samples, the highest toxicity level was evidenced in Al5 (90% effect) followed by OAl1 (79%). The lowest effect was instead measured for Al4bis (16%). By considering the worst case approach, which considers the highest effect obtained in the test battery, apart from the dam, the sampling sites next to the river mouth seemed to be the most impacted ones. A slight effect increase starting from the dam sampling point (OAl1) was also evidenced by bioluminescence inhibition test. Besides, in OAl1 site, although highlightable for all samples, the highest biostimulation (22%, *D. magna*) and also the highest algae growth inhibition were measured. These effects could be attributed to the release from the surrounding areas of agrochemical products acting upon photosynthetic organisms, and to the increased amount of organic substances available as nutrients, given the enclosed area system. The data integration with TBI approach confirmed this observation, highlighting a moderate ecotoxicological risk only for OAl1 (26,3) (Fig. 20.3c). Regarding the sediments, the highest effect was registered by Al1 sample followed by Al5, while the lowest values was obtained for OAl1 with algae slightly inhibited. In the main, it then seemed that the effects were mainly localized in the water matrix and then due to a local temporary discharge. However, a deeper analysis of sediment matrix with a large toxicity test battery should be done to proper evaluate the long term effects also with contact test with whole sediment.

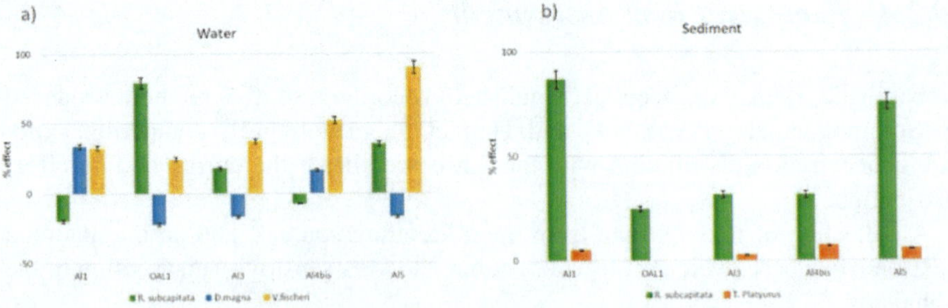

Fig. 20.2 Toxicity (% effect) obtained from the battery test carried out on water (**a**) and sediment (**b**) samples

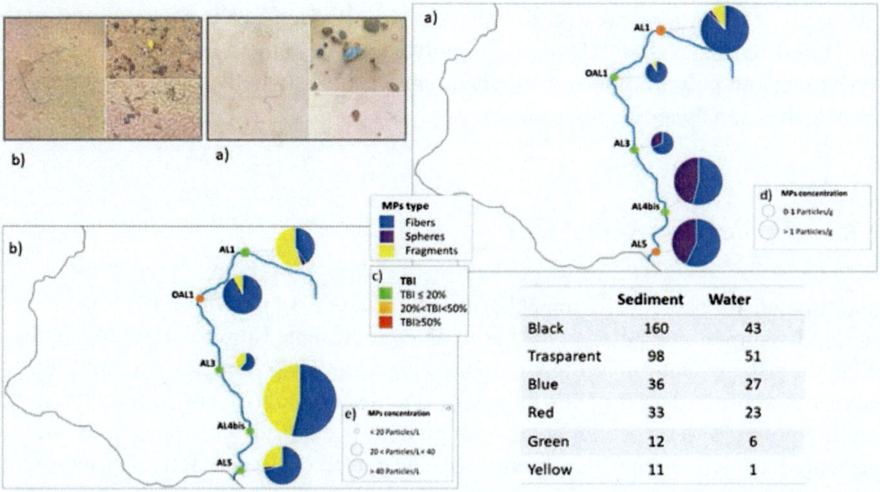

	Sediment	Water
Black	160	43
Trasparent	98	51
Blue	36	27
Red	33	23
Green	12	6
Yellow	11	1

Fig. 20.3 For each sampling site, sediment (**a**) and water (**b**) the TBI (**c**) together with the MPs abundance, morphology and distribution (pie graph) are reported; MP concentrations correspond to the area of the circles particles/g (**d**) and particles/L (**e**). The distribution for colour both for sediment and freshwater are also reported in table

20.3.2 MPs Identification

MPs were identified at all sites both in water and sediment samples. MPs concentration was in the range of 18–43 particles/L and 0.5–1.6 particles/g (Fig. 20.3). For waters, Al4bis sample showed the highest MPs number (43 particles/L). Similarly, in sediment samples, site Al1, Al4bis and Al5 showed the largest amount of MPs with more than one particles/g. Actually, suspended aquatic particulates, including MPs, are expected to be preferentially deposited in low-dynamic environments. In addition, the MPs settlement is largely influenced by polymer type, size, and environmental physicochemical characteristics. Moreover, the presence of tributaries could lead to increased MPs in the final part of the river. However, a clear trend in the MPs concentration from the dam to downstream cannot be evidenced. In the middle tract of the river, we can find that accordance between data relative to ecotoxicological status and MPs occurrence both in water and sediments. It is worth noting that Al3 showed the lowest values for TBI and MPs occurrence.

Fibers were the dominant MPs types both for water and sediment samples followed by fragments for water samples, and spheres for sediment samples (Fig. 20.3). The distribution of plastic colors in the river was similar to other published data in freshwater environments [17]. Black particles were the most common ones, followed by transparent and colored MPs (Fig. 20.3). MPs can inherit their colors from their parent plastic products, but their colors can change due to photodegradation and residence time in the water.

In fishes, MPs fragments and fibers were mainly found in the gastrointestinal tract. Based on the ATR-FTIR analysis, polyethylene, polyethylene terephthalate, polystyrene, and polypropylene, were the main plastic polymers. For environmental matrices, these analyses are still ongoing.

20.3.3 Ecological Risk Assessment

The estimated 5th percentile from MPs toxicity data was 14 part/ L and 91 part/ g for water and sediment, respectively. These values represented the toxicity thresholds used in the risk estimation. In Fig. 20.4, the regression lines characterizing the distributions of exposure data for MPs in water (a) and sediment (b), concerning the five sampling points, and respectively the 95th percentile exposure concentrations, were represented. Moreover, the 5th percentiles of the corresponding toxicity distributions were reported to estimate the risk posed by the MPs to the aquatic organisms in the Alento river.

The MPs concentration 95th percentile for water was 97 part/L, and the probability of exceeding the toxicity threshold (14 part/L) was 35% (Fig. 20.4a). In the case of sediment, the concentration 95th percentile was 7 part/g, and the probability of exceeding the toxicity threshold (91 part/g) was 0.14% (Fig. 20.4b).

These data suggested that in Alento river, the ecological risk posed by MPs on aquatic organisms was not negligible in water, while no risk was assessed for the sediment matrix. These results represented a preliminary assessment of the MPs impact on the Alento River and more detailed studies are needed, especially to expand the set of exposure and toxicity data used in the probabilistic risk assessment.

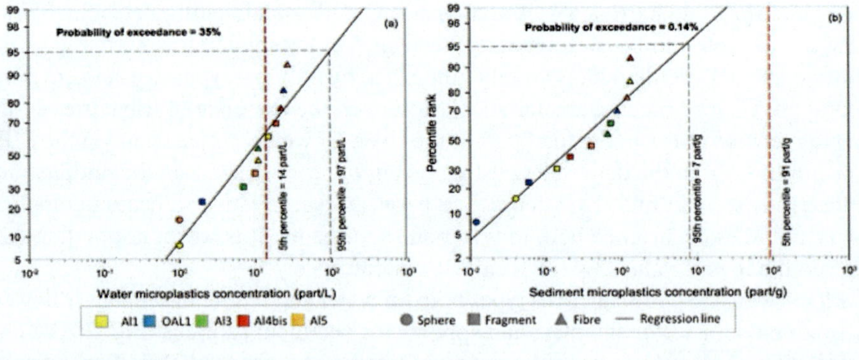

Fig. 20.4 The rank percentiles are plotted on a probability scale against the logarithm of concentrations of microplastics measured in water (**a**) and sediment (**b**) of Alento river, and the linear regression to characterize the corresponding exposure distribution is drawn. The 95th percentile exposure concentration and the probability of exceedance the 5th percentile of the toxicity distribution are represented

Acknowledgements This research is carried out within the project FEAMP: MISURA 1.44—Pesca nelle acque interne e fauna e flora nelle acque interne - art. 44 del Reg. (UE) n. 508/2014. Moreover, this work is also part of the PhD research project of Sara Accardo funded by the University of Naples "Parthenope" and Municipality of Vallo della Lucania and Novi Velia.

References

1. Watkins, Lisa et al. (2019) The effect of dams on river transport of microplastic pollution. Sci Total Environ 664, 834–840
2. Zhang, Bin, et al. (2020) Microplastics in soils: a review of possible sources, analytical methods and ecological impacts. J Chem Technol Biotechn 95.8, 2052–2068
3. Li Lu et al. (2019) Microplastics contamination in different trophic state lakes along the middle and lower reaches of Yangtze River Basin. Environ Poll 254, 112951
4. Dutka AJ, Kochanek PM, Hallenbeck JM (1989) Influence of granulocytopenia on canine cerebral ischemia induced by air embolism. Stroke 20(3):390–395
5. Nendza M (2002) Inventory of marine biotest methods for the evaluation of dredged material and sediments. Chemosphere 48(8):865–883
6. Linea Guida 67/2011 "Batterie di saggi ecotossicologici per sedimenti di acque salate e salmastre"
7. ISO 11348–2:2007(en) Water quality — Determination of the inhibitory effect of water samples on the light emission of Vibrio fischeri (Luminescent bacteria test) — Part 2: Method using liquid-dried bacteria
8. OECD (2011) Test No. 201: Freshwater alga and cyanobacteria, growth inhibition test, OECD guidelines for the testing of chemicals, Section 2, OECD Publishing, Paris
9. OECD (2004) Test No. 202: Daphnia sp. Acute Immobilisation Test, OECD Guidelines for the Testing of Chemicals, Section 2, OECD Publishing, Paris
10. ISO 14380:2011 Water quality. Determination of the acute toxicity to Thamnocephalus platyurus (Crustacea, Anostraca)
11. Manzo, Sonia et al. (2014) Application of a toxicity test battery integrated index for a first screening of the ecotoxicological threat posed by ports and harbors in the southern Adriatic Sea (Italy). Environ Monit Assess 186.11, 7127–7139
12. Alam Firdha Cahya et al. (2019) Microplastic distribution in surface water and sediment river around slum and industrial area (case study: Ciwalengke River, Majalaya district, Indonesia). Chemosphere 224, 637–645
13. Chubarenko I et al (2016) On some physical and dynamical properties of microplastic particles in marine environment. Mar Pollut Bull 108(1–2):105–112
14. Solomon Keith R et al. (1996) Ecological risk assessment of atrazine in North American surface waters. Environ Toxicol Chem Int J 15.1, 31–76
15. Hall Jr, Lenwood W et al. (1999) An ecological risk assessment for the use of Irgarol 1051 as an algaecide for antifoulant paints. Crit Rev Toxicol 29.4, 367
16. Adam V, Yang T, Nowack B (2019) Toward an ecotoxicological risk assessment of microplastics: comparison of available hazard and exposure data in freshwaters. Environ Toxicol Chem 38(2):436–447
17. Hidalgo-Ruz, Valeria, et al. (2012) Microplastics in the marine environment: a review of the methods used for identification and quantification. Environ Sci Technol 46.6, 3060–3075

Chapter 21
Compact Holographic Imaging and Machine Learning for Microfibers Quantification in Laundry Wastewater

Teresa Cacace, Marco del Coco, Mariacristina Cocca, Andouglas Goncalves da Silva Junior, Luiz Marcos Garcia Goncalves, Pierluigi Carcagnì, Melania Paturzo, and Cosimo Distante

21.1 Introduction

Microfiber shedding caused by the washing of synthetic textiles is widely recognized as a major source of primary microplastics in the environment [1]. Recent studies concentrate on the quantification of microplastics released during a wash, also evaluating the influence of different washing treatments or textile characteristics [2–4].

Different methodologies have been employed to detect and identify synthetic microfibers dispersed in water [4]. Often, the filtered fibers are also inspected visually by expert users [2, 3, 5]. This procedure is intrinsically time-consuming and not applicable to continuous monitoring. Moreover, it requires a sample preparation procedure, implemented in a strictly controlled environment to prevent sample contamination [6]. Therefore, there is a need to advance instruments and techniques for fast and automated microparticle identification and counting, while at the same time implementing secure sample handling procedures.

T. Cacace (✉) · M. Paturzo
Institute of Applied Sciences and Intelligent Systems, National Research Council, Pozzuoli, Italy
e-mail: teresa.cacace@isasi.cnr.it

M. del Coco · P. Carcagnì · C. Distante
Institute of Applied Sciences and Intelligent Systems, National Research Council, Lecce, Italy

M. Cocca
Institute for Polymers Composites and Biomaterials, National Research Council, Pozzuoli, Italy

A. G. da Silva Junior · L. M. G. Goncalves
Department of Computer Engineering and Automation, Federai University of Rio Grande Do Norte, Natal, Brazil

M. Cocca et al. (eds.), *Proceedings of the 3rd International Conference on Microplastic Pollution in the Mediterranean Sea*, Springer Water,
https://doi.org/10.1007/978-3-031-34455-8_22

In this framework, we propose a compact and low-cost holographic microscope, able to analyze microplastic samples dispersed in water and flowing through a commercial microchannel. The device implements the architecture for off-axis Digital Holography (DH), which is a non-destructive and single-shot imaging technique. It provides label-free imaging, quantitative phase mapping and flexible a-posteriori numerical refocusing of flowing samples. At the same time, the use of microfluidics for fluid handling provides the basis for high-throughput and automatic acquisitions, while reducing the risks of sample contamination.

We employ our device to analyze an aliquot of wastewater from a washing test of polyester textiles in a commercial washing machine. Moreover, we propose an ad-hoc data processing procedure to estimate the microplastic concentration, based on image processing and machine learning techniques.

21.2 Experimental

21.2.1 Data Acquisition

We have recently introduced a compact and portable DH design (Fig. 21.1), which implements off-axis DH, an imaging technique based on the recording of the interference between the light scattered by the sample (object wave) and a clean reference wave. The resulting hologram is recorded by a digital sensor and then numerically processed to yield amplitude images and quantitative phase maps of the sample, and information on the size, shape, and tridimensional position of the recorded objects. The samples do not require any pre-treatment or staining and are not damaged during the image acquisition. Moreover, off-axis DH relies on single-shot measurements, enabling dynamic image acquisition of flowing samples for high-throughput analysis.

The microscope design implements a wavefront division interferometric configuration [7, 8] to acquire the robustness against vibrations necessary for in-situ measurements. For the light source, we use a compact and low-cost DPSS laser (Thorlabs, CPS532), which emits a collimated beam at $\lambda = 532$ nm. Then, a neutral filter reduces the light intensity on the camera, avoiding pixel over-exposure. The light impinges orthogonally on the sample flowing through a commercial microfluidic straight-channel chip (PMMA, $1.0 \times 0.2 \times 58.0$ mm) and is then collected by a compact microscope objective (Newport, 10x). Afterwards, a commercial grating (Thorlabs, 300ln/mm) collects the section of light which passes through the microfluidic platform, immediately adjacent to the channel and free from the object signal. The light is tilted by the grating and then collected on the camera (See3CAM_CU55M). In this way, we reproduce the off-axis configuration by recording the interference between the object wave and the tilted reference wave. Moreover, the integration of the optical components in a single monolithic block greatly reduces the influence of vibrations on the system. Note that it is possible to change the microfluidic chip when needed, to avoid cross-contamination between samples.

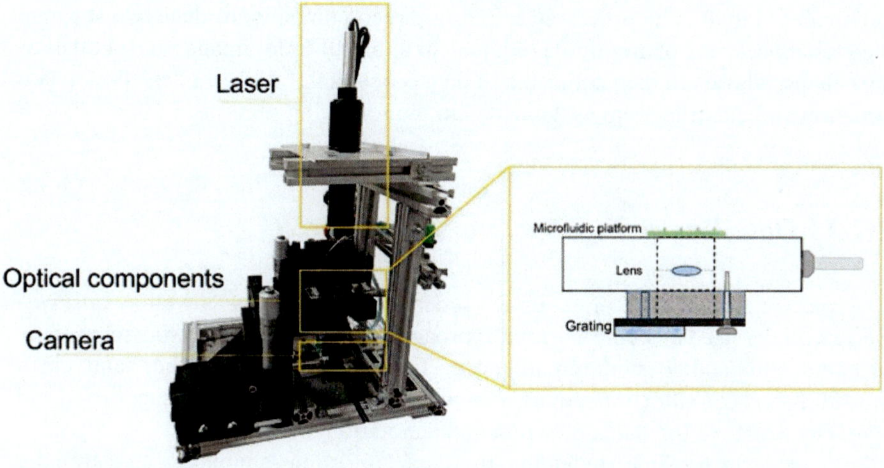

Fig. 21.1 Compact holographic microscope. All optical components are integrated into a single monolithic block to limit the effect of external vibrations

The proposed holographic microscope has been recently used for the imaging of flowing microplastics, demonstrating the reconstruction of amplitude images and phase maps and the numerical refocusing of out-of-focus objects [8]. Here, we use it to study a sample of wastewater obtained from the washing process of synthetic fabric performed at real scale. The washing test is performed using a Bosch washing machine serie 4 varioperfect WLG24225i, at 40 °C, for 107 min and 1200 rpm, using a commercial liquid detergent, in the recommended doses, and a washing load of 2–2.5 kg made of 100% polyester t-shirts. The wastewater coming directly from the drainpipe of the washing machine subjected to a multi-step filtration procedure. For this study, we have selected an aliquot of wastewater collected after passing through a nylon net filter with a 20 μm pore size (Merck millipore). Moreover, in order to increase the density of microplastics and speed up the analysis process, a rotary evaporator is employed to halve the sample volume by evaporation under reduced pressure. Note that, while the water pre-treatment procedure has been adopted to restrict the scope of the work and speed up the data acquisition process, there is no impediment to the use of untreated wastewater. The only limitation is the size of the microfluidic channel, which should be adjusted to prevent clogging.

The samples are kept in flow in the microchannel using a pressure-driven pump. We have acquired video recordings of the flowing samples, using FHD resolution (1920 × 1080 pixels, pixel size = 2.2um) and a framerate of 10 fps. The width of the channel (1 mm) is recorded entirely by using a low magnification of about 4x. The collected videos are numerically processed frame-by-frame, according to a standard procedure for hologram demodulation. The latter is based on linear filtering in the Fourier domain, and it retrieves from each frame an amplitude image and the corresponding quantitative phase map [8]. Moreover, optical aberrations are removed by numerical subtraction of a reference hologram, which is obtained as the mean over the whole video. Finally, the processed frames are recombined to produce two separate video sequences, one of amplitude (equivalent to bright-field) images and another of phase maps. Those are then subjected to data processing for object detection, patch extraction and classification, as described in the following sections.

21.2.2 Data Processing Pipeline

The processing pipeline exploits image processing techniques as well as unsupervised and supervised machine learning approaches. The image processing regards the object detection and does not require any preliminary step. On the other hand, patch embedding and classification require a training phase (respectively unsupervised and supervised) making the data collection and annotation mandatory.

The complete pipeline, including the data collection, annotation and training phases, is represented in Fig. 21.2. The first row shows the flow of the main pipeline where the phase video sequence is provided, frame by frame, to the *object detection* module. The detected objects are then processed by the *embedding* module that exploits the trained model in order to return a 32-dimensional vector representative of the patch given as input. Such a vector is then sent to the *classification* module that returns the predicted class.

Both embedding and classification modules require a trained model, respectively *Latent model* and *SVM model*. The generation of the latent model is based on an unsupervised approach that requires a sample set of patch representative of both

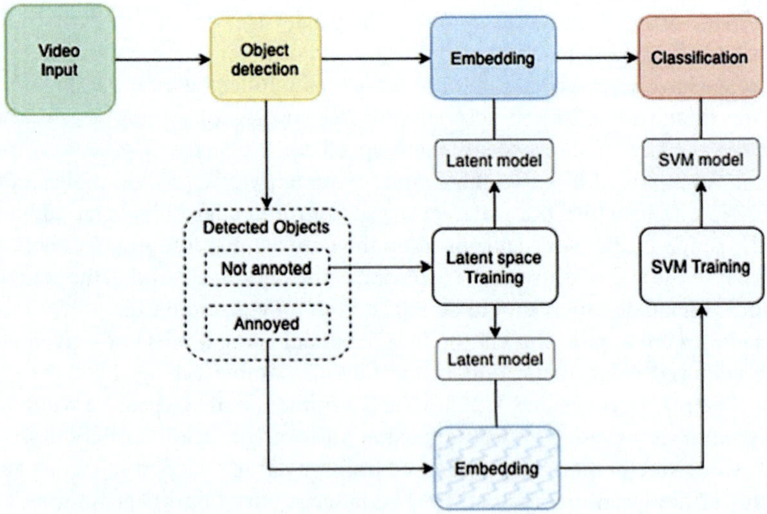

Fig. 21.2 Data processing pipeline: the upper part of the image shows the main pipeline, the lower part shows the training phases

the classes, but no specific annotation. On the other hand, in order to compute the SVM model, a well-annotated set of samples is mandatory. Thus, the training phase starts by randomly splitting the detected objects into two subsets: the first one is used as it is for the training of the latent model; the other is manually annotated as "microplastics/non-microplastic" by a human operator, processed by the embedding module to generate the pair patch-embedding/class and finally employed for the SVM model training.

Object detection is based on the Mixture Gaussian Model [9] foreground detector. The algorithm discriminates between foreground and background pixels by means of a continuously updated model based on a mixture of gaussians representative of a non-stationary background. The pixels that do not fit the model are considered to be foreground. This kind of approach is particularly suitable for the proposed application where the background is subject to slight fluctuation due to residual vibrations during the recording.

Once the foreground pixels have been detected, connected components labelling is used to group them into objects, which are finally defined by a bounding box. Overlapping bounding boxes are merged together to avoid object fragmentation caused by misdetection.

Embedding module exploits the idea of a latent space generated by an *autoencoder neural network* [10]. The key idea is that high dimensional data (e.g. the patch image) has often a lower dimension embedding which is sufficient to represent all the informative content of the original data. Such an embedding led the extracted dataset of patch in a *Latent space*, a lower dimensional manifold of the high dimensional images where we expect all the instances of the dataset to lie in proximity. The desired embedding is obtained by means of an autoencoder neural network, a kind of symmetrical (encoder/decoder) neural network where the output is the image itself as represented in Fig. 21.3. The training aims at the optimization of the network weight so that the image provided in output is as close as possible to the input one. This optimization intuitively led to a latent space embedding all the information necessary for the image reconstruction and then a space that is highly representative of the input. Once the network is trained, the encoder side will be used to accomplish the embedding processing.

Classification module is based on a Support Vector Machine SVM classifier [11]. The main idea behind the algorithm is to find the hyperplane that optimally separates the training sample depending on their classes. More precisely a support vector classifier has been employed.

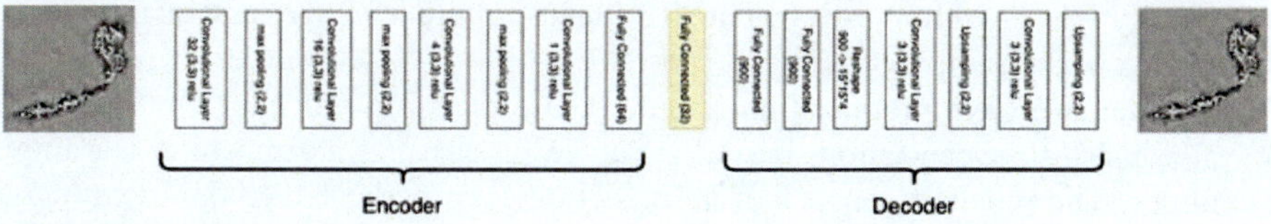

Fig. 21.3 Autoencoder network

21.3 Results and Discussion

The data processing has been performed on a video sequence of 4240 frames, from which the object detector has retrieved a dataset of 5560 patches. Among them, 2608 patches have been annotated with a specific class (577 microplastic and 1758 non-microplastic) and exploited to train the classification model. The other unannotated 2952 patches have been exploited for the unsupervised learning required by the autoencoder training. This choice guarantees that the embedding generated with the annotated patches is performed over unseen samples.

The autoencoder employed in the embedding module has been trained, as stated before, over the 2952 unannotated patches. In order to increase the generalization capabilities and to contrast the scarcity of the "microplastic" objects, a data augmentation has been performed by means random rotation and translation of the given patches over a total of 15 epoch performed with a batch size of 64 elements.

In order to return a measure of the classification accuracy, we worked on the classification step in an leave-one-out fashion. Given the embedding of the annotated patches and their annotations (labels) the classifier has been iteratively trained on n-1 samples (where n is the total number of samples) and tested on the sample kept aside. Such a procedure has been repeated for different values of the SVM hyperparameter in order to find the optimum value.

The best accuracy led to the following results:

[330 147]

[268 1490]

[0.692, 0.308]

[0.152, 0.848]

That led to an overall accuracy of 77%, a non-negligible result considering the 1 to 10 class unbalancing in the employed training set.

21.4 Conclusions

We have presented a compact and low-cost holographic microscope, using it to detect microplastic samples dispersed in laundry wastewater from a commercial washing machine. The combination of dynamic image acquisition and microfluidics for sample handling provides the basis for high-throughput and automatic acquisitions. We have recorded holographic videos of the flowing sample. They have been numerically processed frame-by-frame to retrieve amplitude images and quantitative phase maps of the sample.

Moreover, we have proposed an ad-hoc data processing procedure to estimate the microplastic concentration, based on image processing and machine learning techniques. The pipeline consists of an object detector dedicated to the identification of the mobile objects in video sequences, a deep learning module devoted to projecting the detected objects into a low dimensional space and finally a classification step responsible for the classification in one of the two categories "microplastic"/ "non-microplastic" that enables the microplastic density estimation in a given water sample. Behind the scenes, the development of the mentioned approach required specific data annotation and training phases. Despite the data unbalancing, we obtained an object classification accuracy of 77%.

References

1. Boucher J, Friot D (2017) Primary Microplastics in the oceans: a global evaluation of sources. IUCN, Gland, Switzerland, p 43
2. De Falco F et al (2019) The contribution of washing processes of synthetic clothes to microplastic pollution. Sci Rep 9:6633
3. Volgare M et al (2021) Washing load influences the microplastic release from polyester fabrics by affecting wettability and mechanical stress. Sci Rep 11:19479
4. Gaylarde C et al (2021) Plastic microfibre pollution: how important is clothes' laundering? Heliyon 7(5):e07105
5. Corami F et al (2020) A novel method for purification, quantitative analysis and characterization of microplastic fibers using Micro-FTIR. Chemosphere 238:124564
6. Masura J et al. (2015) Laboratory methods for the analysis of microplastics in the marine environment: recommendations for quantifying synthetic particles in waters and sediments. Siver Spring, MD, NOAA Marine Debris Division, p 31 (NOAA Technical Memorandum NOS-OR&R-48)
7. Cacace T et al (2020) Compact off-axis holographic slide microscope: design guidelines. Biomed Opt Express 11:2511–2532
8. Cacace T et al. (2021) Compact holographic microscope for imaging flowing microplastics. 2021 International Workshop on Metrology for the Sea - Learning to Measure Sea Health Parameters, pp 229–233
9. Zivkovic Z (2004) Improved adaptive gaussian mixture model for background subtraction. Pattern Recognition, 2004. ICPR 2004. Proceedings of the 17th International Conference on 2, pp 28–31. IEEE
10. Kramer, Mark A (1991) Nonlinear principal component analysis using auto associative neural networks. AIChE J 37, 2, 233–243
11. Chang CC, Lin CJ (2011) LIBSVM: a library for support vector machines. ACM Trans Intell Syst Technol (TIST) 2(3):1–27

Chapter 22
Microplastic in Drinking Water: A Pilot Study

Eleonora Brancaleone, Daniela Mattei, Valentina Fuscoletti, Luca Lucentini, Gabriele Favero, Alessandro Frugis, Valentina Gioia, and Marco Lazzazzara

22.1 Introduction

In recent years, microplastic pollution has become increasingly important in the eyes of public opinion. For years, plastics have been used in various production contexts, such as health, food or technology, thanks to both their chemical-physical properties and their 'shelf-life', which makes them almost indispensable products in everyday life.

While these properties have made the use of plastics advantageous in various fields, they also have a negative impact on the environment and, consequently, perhaps on animal and human health.

A relevant example is the impact that these materials seem to have on the availability and quality of water resources, affecting all productive sectors, in particular primary production and food production, sanitation, human and animal safety use.

This impact represents a real water emergency also because microplastics are ubiquitous emerging contaminants with a still unclear toxicological profile. In fact, despite the media interest on the subject, there is no complete and exhaustive picture of the toxicity of microplastics on humans.

Therefore, based on the provisions of the Directive (EU) 2020/2184, it is fundamental to develop monitoring and prevention strategies centred on site-specific risk analysis in accordance with the provisions of the Water Safety Plans, to guarantee

E. Brancaleone · D. Mattei · V. Fuscoletti · L. Lucentini
Department of Environment and Health, Italian National Institute of Health (ISS), Rome, Italy

E. Brancaleone · G. Favero
Department of Environmental Biology, Sapienza University of Rome, Rome, Italy

A. Frugis · V. Gioia · M. Lazzazzara (✉)
Acea Elabori SpA, Rome, Italy
e-mail: marco.lazzazzara@aceaspa.it

M. Cocca et al. (eds.), *Proceedings of the 3rd International Conference on Microplastic Pollution in the Mediterranean Sea*, Springer Water,
https://doi.org/10.1007/978-3-031-34455-8_23

the monitoring of water resources and, therefore, prevent this state of emergency caused by parameters that are not currently provided for by law [1, 2].

Therefore, the prospect of being able to develop a monitoring strategy for these contaminants is important to reduce potential exposure and prevent possible health effects.

The European Commission has set itself the goal of regulating the monitoring of these emerging contaminants by 2024, identifying a method for their determination and characterisation, to enable their inclusion in the Watch List and reduce potential exposure to these substances through water.

Therefore, the prospect of being able to develop microplastic monitoring and prevention strategies is important to assess potential exposure and prevent and/or reduce any confirmed risk of potential exposure to this type of contaminant.

22.2 Experimental

Water samples from different sample points along the drinking water chain were analyzed. Analyses were carried out on raw and treated water samples and, subsequently, on drinking water sampled from the drinking fountain, the home tap and dispense water boxes to assess the actual consumer exposure to these contaminants from different points of supply.

22.2.1 Materials

In order to be able to assess possible consumer exposure to these contaminants, the quality of drinking water sampled at various points in the drinking water supply chain was monitored by Raman Spectroscopy analysis. In this context, for optimizing efficiency and analysis time, the drinking water samples under investigation are pretreated with Hydrochloric Acid (by Carlo Erba) to reduce the high background noise due to inorganic substances and other solids and, therefore, facilitate analysis by μ-Raman. They are, then, filtered through a special filtration system, supplied by Thermo Fisher Scientific, consisting of a porous septum in which the silicon filter is placed (pore size 1 μm and 10 × 10 mm of laser dicing) and closed by a black gasket to prevent sample leakage and a flushing apparatus; filtration volume varied between 250–500 mL.

Counting and identification of particles was performed with a μ-Raman Spectroscope (Thermo Fisher Scientific) equipped with three objectives, specifically 10x, 20x and 50x and a 532 nm Raman laser (0,1–10 mW; 2 cm^{-1} spectral resolution) with adjustable laser power investigating microplastic size range upper 20 μm.

22.2.2 *Methods*

The Raman effect is based on inelastic light scattering of sample chemical bonds; due to vibrations in the chemical bonds, this interaction causes a specific energy shift in parts of the back scattered light which results in a specific Raman spectrum.

Raman scattering can be differentiated as Stokes and Anti-Stokes scattering, which both contain information about the sample and its molecular composition. In addition, elastic scattering, called Rayleigh scattering, can also appear and it has the same energy as the incident radiation, but it does not contain any information about the molecules and cannot be used for chemical sample analysis in a confocal Raman microscope.

Therefore, Raman spectrum is the fingerprint of a sample, from which qualitative and quantitative information can be obtained.

The confocal Raman imaging technique combines Raman spectroscopy with a confocal microscope, capturing the information of a complete Raman spectrum in each pixel of the image. In this way, the spatial distribution of the particles of interest within the sample is detected and photographed.

For analysis with Raman imaging, a confocal Raman microscope scans the sample point by point and line by line, and a full Raman spectrum is acquired at each pixel of the image (hyperspectral imaging). Spectra, containing chemical information, are analyzed with Omnic Spectra Atlus software to generate an image that visualizes the spatial distribution of the particles of interest. This equipment provided easy and quick localization, counting and morphological characterization (size, area, length, width) of the particles presented on the filter.

After particle localization and morphological characterization, automated Raman analysis was performed: spectrum was collected on the center of each particle; parameters were selected to obtain a good compromise between spectra quality and time needed for analysis. The chemical identification of particles was realized using commercial Raman libraries (e.g. HR FT-Raman Polymer Library, Polymers & Additives_Packaging) and a homemade library.

22.2.2.1 Preparation of Samples

To reduce background noise, all samples were pretreated, at room temperature, with HCl (pH = 3) and then each one was filtered through the vacuum filtration system. Recovery of any microplastic adhered to the sampler or flussex was performed using Milli-Q water. Once the vacuum was broken, the filter was placed under μ-Raman's microscope for subsequent analysis.

22.2.2.2 Analytical Techniques

In order to enhance and optimize the contrast between the substrate and particles for their qualitative-quantitative identification, it was decided to work in the dark field at tenfold magnification to get an overall picture of the sample under examination; in this way, it is possible to clearly distinguish the particles under examination from the dark background given by the silicon filter, thus ensuring that any loss of particles of interest is avoided.

Therefore, after placing the filter being analyzed under the μ-Raman's microscope, a single microscope image is acquired to have a total mapping of the particles present over the entire filter area and obtain the information of interest for the determination and characterization of microplastics in drinking water samples.

Image analysis allows the evaluation of particle number, size and morphology in the size range of interest, specifically above 20 μm. The spectrum of each particle was compared to reference libraries, appropriately chosen for the identification of microplastics: the latter grouped a wide range of the most common polymers and commercial materials.

The library match was further extended with additional spectra obtained from materials either specifically used for this type of analysis or commonly found in the laboratory that were plastics. Library matches with a match greater than 70% were accepted; conversely, library matches with a match between 60 and 70% were analyzed and interpreted individually by overlaying the spectra.

22.3 Results and Discussion

In total, 30 samples of water for human consumption taken at different points in the drinking water chain were analyzed. Five different drinking water matrices were analyzed, specifically, 14 samples of groundwater, both pre and post treatment of the drinking water purifier, 8 samples of drinking water taken from dispenser boxes and a final batch of samples taken from a drinking fountain and domestic tap. All results were normalized to the 1 L volume of the sample.

22.3.1 Results

The batch of groundwater samples consisted of 4 samples taken at the inlet of the drinking water treatment plant (Groundwater-I) and 10 post treatment (Groundwater-O): on average, the incoming water contained 5 ± 1.5 particles/L, as shown in Fig. 22.1. Polymer identification revealed the following 15% of the particles were PET, 10% PTFE while the remaining 75% is attributed to other suspended particles not identifiable as polymeric materials but due to inorganic compounds or other solids (Fig. 22.2).

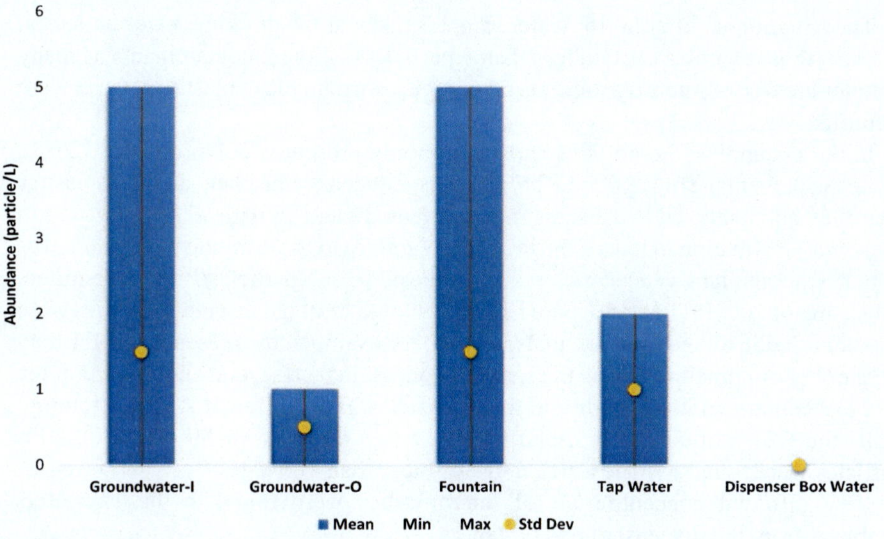

Fig. 22.1 Average Content of Microplastic in water of different types

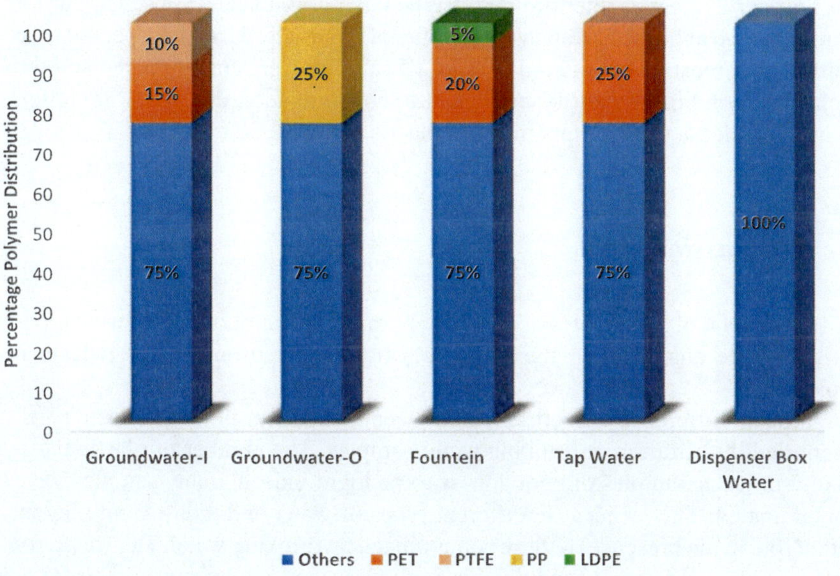

Fig. 22.2 Polymer distribution of the microparticles found in water

The situation is different for water samples taken at the drinking water outlet: on average, these samples contain less than 1 particle/L of these contaminants as many samples are free of microplastics, i.e. no particles attributable to plastic material were identified.

In these samples, the polymer most commonly present is polypropylene (25%), while the percentage of PET and PTFE input is reduced by the plant; all microplastics identified and analyzed had a size range between 30 and 90 μm.

A similar situation to that of the incoming samples to the drinking water plant was found in the fountain water samples: an average of 5 ± 1.5 particles/L were identified, consisting of 20% PET and 5% pure LDPE. It should be noted that microplastics were not identified in all the samples analyzed and, for example, the presence of PET may be due to contamination during the sampling phase; in fact, several aliquots were free of these contaminants. In addition, these particles had a different size distribution: while the PET particles were found to have a size range of 30–90 μm, the LDPE particles were found to have a size range of more than 100 μm.

Very different concentrations of microplastics were found in the 12 waters analyzed from the dispenser box and tap.

In fact, these samples have a lower content of microplastics: specifically, the water from the dispenser box is completely free of these pollutants, showing, in terms of particle number/L, a better filter profile than the laboratory blanc. Conversely, water taken from the tap at home had an average value of 2 particles/L, in this specific case PET, although it must be considered that not all samples were contaminated and the microplastics found can be attributed, again, to contamination during the sampling phase and the release of tap filters made of plastic material.

22.3.2 Discussion

In this study a total of 30 samples of water for human consumption of different types and origins were analyzed for their microplastic content by means of μ-Raman spectroscopy.

A significant difference was observed between the number of microplastic particles identified in the different drinking water samples. The number of plastic particles found in water samples did not appear to be high; indeed, many samples were not contaminated. This is because different types of water and their origin play an important role in the presence of these contaminants in drinking water. Tap water, for example, is obtained from different sources of raw water, e.g. from groundwater or surface water. The latter may be contaminated with varying amounts of microplastics (MPs): groundwater, due to the filtering action of the soil, may contain less MPs than groundwater. Conversely, raw water from surface water may have a higher MPs content because it is directly exposed to the environment. Added to this is the fact that, before being supplied as tap water, groundwater is treated through different stages, depending on its initial conditions, such as coagulation/flocculation, sedimentation, filtration, and disinfection. Furthermore, it must be considered that the removal efficiency also depends on the size of the MPs analyzed. In addition to MPs reduction, drinking water treatment steps can also increase the number of MPs: polymeric membranes used during filtration processes may leach MPs acting as a source of contamination. Finally, in addition to treatment processes, it must be considered that drinking water is usually delivered through a pipe system consisting of different materials, including plastic, which could lead to (re)contamination [3].

To confirm this, it can be observed that the chemical composition of these polymers is similar; the polymers identified in the various matrices are PET, PTFE, PP or PE, the most common plastics and present in plants and piping systems. To this should be added that a high risk of contamination may also derive from the sampling phase, in fact, PET was identified in four out of five matrices and is assumed to be due to the glass bottle caps used to sample.

22.4 Conclusions

The assessment of microplastics in drinking water is a multi-step process, from sampling to quantification of microplastics, with each step requiring time and attention to possible sources of contamination. The high variability of the chemical composition and morphology of microplastics is a crucial point for monitoring these contaminants to make a possible assessment of consumer exposure through drinking water. The present study offers a reproducible and effective method for the determination and characterization of microplastics in water for human consumption using μ-Raman Spectroscopy. The rapid analysis of samples makes it possible to assess probable consumer intake by examining drinking water supplied from various points in the supply chain and, if necessary, try to identify the possible source of contamination.

Therefore, μ-Raman Spectroscopy can be successfully used for the detection and analysis of microplastics (>20 μm) in water for human consumption: the high resolution of μ-Raman Spectroscopy is required to detect the presence of particles of interest below a certain size range.

In conclusion, this pilot study could be a suitable routine method for monitoring microplastics in drinking water and ensure consumer health protection.

References

1. Directive (EU) 2020/2184 of the European Parliament and of the Council of 16 December 2020 on the quality of water intended for human consumption (recast). Official Journal of the European Union L 435/1, 23 December 2020
2. Martellone L, Lucentini L, Mattei D, De Vincenzo M, Favero G, Bogialli S, Litti L, Meneghetti M, Corami F (2021) Rosso, B: Strategie di campionamento di microplastiche negli ambienti acquatici e metodi di pretrattamento (Rapporti ISTISAN 21/2). Istituto Superiore di Sanità, Rome
3. Pivokonsky M, Cermakova L, Novotna K, Peer P, Cajthaml T, Janda V (2018) Occurrence of microplastics in raw and treated drinking water. Sci Total Environ 643:1644–1651

Chapter 23
Microplastics and Associated Plasticizers: Presence and Detection in Cnidarians Used as Possible Bioindicators for Microplastic Contamination in Marine Environments

Sara Vencato, Francesco Saliu, Valerio Isa, Simone Montano, Davide Seveso, Paolo Galli, Silvia Lavorano, Stefania Coppa, Andrea Camedda, Giorgio Massaro, and Giuseppe Andrea de Lucia

23.1 Introduction

The pollution of the seas caused by the dispersion of plastic litter is one of the most serious environmental emergencies worldwide. Different sampling methodologies have been used to document microplastics presence (<5 mm in size, MPs) in sea water. Current approaches include the use of marine organisms as bioindicators and the detection of plastic-associated contaminants in their tissues, knowing that microplastic can function as vector of contaminants in various organisms.

Between such contaminants, phthalate esters (PAEs) are plastic additives commonly blended with plastic polymers in high relative mass amounts (up to 60% of the total plastic product weight) [1], used to enhance the longevity of plastic materials. Since they are not covalently bound to the plastic polymers, they can easily leach into the environments and become ubiquitous and bioavailable to different marine organisms due to their lipophilicity. However, once absorbed by an organism, they do not bioaccumulate, instead are rapidly metabolized. Indeed, laboratory and field

S. Vencato (✉) · S. Coppa · A. Camedda · G. Massaro · G. A. de Lucia
Institute of Anthropic Impact and Sustainability in Marine Environment, National Research Council, Oristano, Italy
e-mail: sara.vencato@ias.cnr.it; s.vencato@campus.unimib.it

S. Vencato · F. Saliu · V. Isa · S. Montano · D. Seveso · P. Galli
Earth and Environmental Science Department, University of Milano Bicocca, Milano, Italy

S. Lavorano
Costa Edutainment SpA – Acquario di Genova, Genova, Italy

M. Cocca et al. (eds.), *Proceedings of the 3rd International Conference on Microplastic Pollution in the Mediterranean Sea*, Springer Water,
https://doi.org/10.1007/978-3-031-34455-8_24

studies indicate that phthalates do not biomagnify in aquatic food-webs and that higher molecular weight phthalates show evidence of trophic dilution in aquatic food-webs, which indicate that metabolic transformation is a key mitigating factor [2].

Phthalates occurrence has already been reported in zooplankton [3], marine invertebrates [4] and marine mammals [5]. At the same time, a possible correlation between MPs exposure and PAEs presence was proposed in different marine organisms [4, 5]. Consequently, the presence of phthalates was proposed as marker to evaluate MPs contamination of marine environments. Thus, phthalates may be considered a short range indicator of the interaction between marine organisms and plastic litter and allow to appreciate the effectiveness of potential mitigation measures in the short term.

However, on this topic there is a lack of data, mostly due to analytical difficulties. Indeed, background contamination and the ubiquity of plasticizers in both laboratory and natural environments may interfere during sample collection and analyses [6, 7].

Recently, a procedure involving solid phase microextraction (SPME) coupled with liquid chromatography mass spectrometry (LC/MS) analysis has been proposed as a non-lethal alternative (bioSPME-LC–MS/MS) in order to quantify phthalate esters in marine organisms, offering at the same time an improved control of the background contamination compared to the classical extraction procedure [6]. For the best of our knowledge, there is no data regarding rates of direct transfer of PAEs into cnidarian tissues based on microplastic exposure. Jellyfishes were reported as target organisms for marine litter, being able to internalise both macro- and microplastics [8]. Experiments show that scleractinian corals can ingest microplastics suspended in the water [9]. Octocorals and sea anemones are benthic soft-cnidarians with a worldwide distribution that share different physical traits, have similar ecological roles [10] but have tended to be overlooked [11]. Still, octocorals offer different services that underpin ecosystem biodiversity [10]. Sea anemones, as opportunistic feeders, may be particularly affected by microplastics consumption, which makes them an excellent potential target to monitor microplastics contamination [12]. However, studies on plastic debris consumption by anthozoans are still scarce. Therefore, at the Genoa Aquarium facilities we investigated the capacity of a soft coral (*Coelogorgia palmosa,* Milne-Edwards & Haime, 1857) to interact with MPs at different microplastics experimental concentrations. Then, we measured the presence and the bioconcentration factor (BCF) of 8 common PAEs in different soft coral species raised in the same microcosm environment, using the bioSPME-LC–MS/MS procedure. BCF are used to evaluate the inclination of aquatic organisms to accumulate chemicals from their ambient environment [13].

Main goal of this work is to investigate the possible use of PAEs as a marker to evaluate microplastic contamination in the marine environments. So, we propose the application of the bioSPME-LC–MS/MS technique using cnidarians as bioindicators of PAEs presence. In order to test the validity and sensibility of such methodology, we aim to test it at very different environmental conditions, both in the field and in laboratory and with different soft-benthic cnidarian species.

Indeed, knowing that, on-field, there are different MPs concentrations and PAEs levels are extremely variable in terms of space, time and plastic conditions, we aim also to evaluate the role of MPs in transferring PAEs into marine organisms at environmentally detected concentrations. To this end, a study on the assessment of sea anemones of the species *Anemonia viridis* (Forsskål, 1775) and *Actinia equina* (Linnaeus, 1758) as target organisms for monitoring the PAEs presence is taking place in the waters around the Sinis Peninsula (Gulf of Oristano, Sardinia). MPs and PAEs presence in both the species will be assessed in an area where the abundance and distribution of microplastics and phthalates have already been quantified in seawater [14].

23.2 Experimental

At the Genoa Aquarium facilities, *Coelogorgia palmosa* was chosen to explore the interaction mechanisms between microplastics and octocorals. The determination of the phthalate concentration levels in coral tissue was performed by employing bioSPME-LC–MS/MS [6] on the soft corals *Coelogorgia palmosa, Sinularia sp., Sarcophyton glaucum,* and *Lobophytum sp.*

For the "on site" counterpart of the work, specimens of *Anemonia viridis* and *Actinia equina* were randomly collected in 4 different sites around the Sinis Peninsula. Analyses are currently taking place for the evaluation of microplastics and phthalate presence in such organisms through, respectively, MPs visual and polymeric inspection and bioSPME-LC–MS/MS analyses for phthalates characterization.

23.2.1 Materials

13 *Coelogorgia palmosa* fragments were collected from 6 random colonies raised in the aquarium tanks. After the acclimation period, each fragment was transferred into an individual interaction chamber for 48 h (Fig. 23.1). Each interaction chamber was equipped with an air pump, to allow the MPs circulation [15] and allocated in a water bath aquarium's tank to maintain the temperature of 25 °C. Fragments were randomly assigned to 2 treatments: T1 (MPs concentration, 0.01 g/L) and T2 (MPs concentration, 0.07 g/L) with fluorescent PE microbeads (0.98 g cm–3), size range of 180 to 212 μm (Cospheric).

For each treatment, one chamber without the coral (blank) was set up to evaluate the loss of microplastic in the system. Moreover, 3 chambers, each one with a coral fragment but without PE, were used as controls to check the coral health status under experimental conditions [15].

For the study in the field, 40 specimens of sea anemones (20 *Actinia equina* and 20 *Anemonia viridis*) were randomly collected in 4 sites located around the

Fig. 23.1 Interaction chambers system laying in water-bath inside the aquarium water tank

Sinis Peninsula (Western Mediterranean Sea), where both the species could be found simultaneously.

23.2.2 Methods

At the end of the experiment, coral fragments were removed from their chambers and rinsed with salt water to count the number of microbeads adhered under a stereomicroscope (Paralux) integrated with a UV light kit (NIGHTSEA). Then, each coral fragment was placed in a Petri dish for the tissues digestion in sodium hypochlorite [15]. Microbeads were considered as 'adhered' when they were found attached to the coral surface (Fig. 23.2a), while they were considered as 'ingested' when found inside the polyps' mouths (Fig. 23.2b) or observed in the Petri dishes after the complete dissolution of each fragment [16]. Sea anemones samples were treated with H_2O_2 digestion coupled with visual sorting at stereoscopic microscope (Carl Zeiss Microimaging GmbH) equipped with image analysis system (AxioCam ERc5s and Zen, 2014 Blue edition software) for microplastics extraction.

Determination of dimethyl phthalate (DMP) diethyl phthalate (DEP) dibutyl phthalate (DBP) butyl benzyl phthalate (BBzP), di-(2-ethy hexyl) phthalates (DEHP), mono-butyl phthalate (MBP), mono-benzyl phthalate (MBzP), mono-(2-ethy hexyl) phthalate (MEHP) concentration in coral tissues was performed by submitting 5 replicates for each coral species to the bioSPME-LC–MS/MS analysis. The same procedure will be soon performed for the PAEs determination in sea anemones samples.

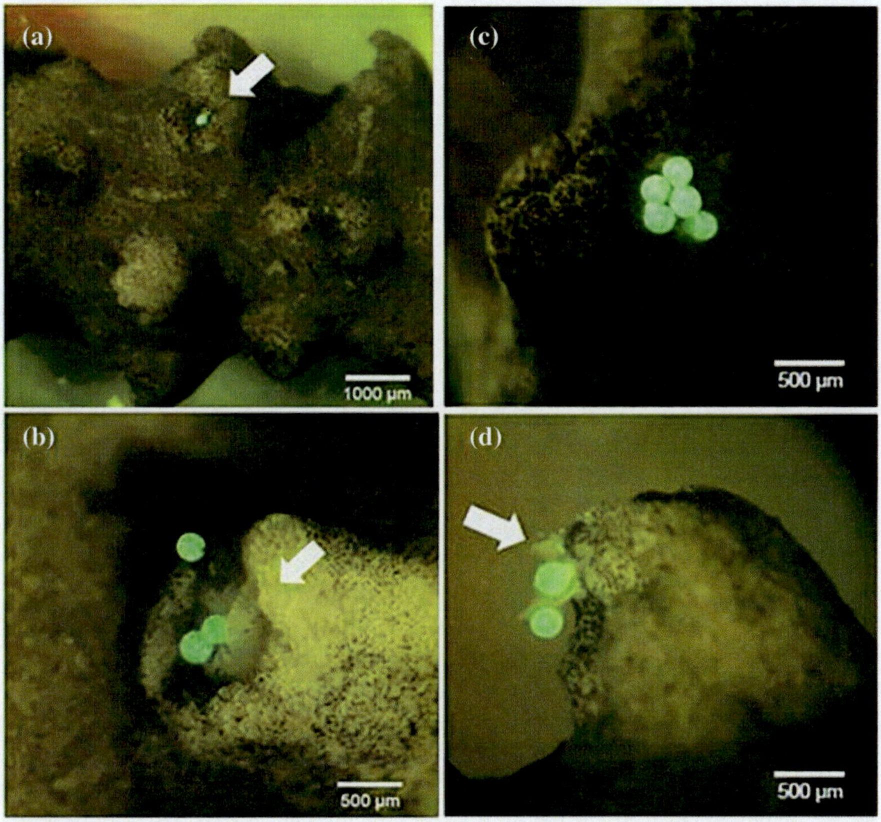

Fig. 23.2 Microplastic interactions with *Coelogorgia palmosa*: **a** a PE bead inside the polyp's mouth, **b** beads trapped inside a polyp: the white arrow shows a polyp tentacle interacting with a bead, **c** adhered polyethylene beads on the coral surface next to a polyp mouth, **d** PE beads trapped by coral mucus: the white arrow shows abnormal mucus coming from a coral polyp (Modified from [16])

23.3 Results and Discussion

23.3.1 Results

Results listed here are preliminary results that mainly described the laboratory already performed part of this study. The on field and Mediterranean related work is still in progress and will be soon finalised.

At the end of the lab experiment, all *Coleogorgia palmosa* fragments showed microbeads stuck to their surface (Fig. 23.2c) and trapped by the produced mucus (Fig. 23.2d). Moreover, microplastics presence in coral polyps and sea anemones was highlighted in both the lab treatments and in nature respectively. This prove the presence of interactions between microplastics and these organisms.

PAEs are absorbed by soft corals tissues, and bioSPME-LC–MS/MS method is able to detect them in such samples under controlled conditions.

23.3.1.1 Microplastic Interactions

The highest adhesion value of PE beads per coral fragment was observed in T2, with an average value 9 times higher than that in T1. Differences in microplastic adhesion between different PE concentrations were not statistically significant ($U = 10$, $z = 0.584$, $p = 0.686$). By contrast, both T1 and T2 showed a statistically significant strong positive correlation between abnormal mucus presence and adhered microplastic number (Kendall's tau-b correlation test, $\tau b = 0.550$, $p = 0.016$). *Coleogorgia palmosa* in T2 reported the highest values of ingested PE beads per coral fragment but no statistically significant differences in microplastic ingestion between the treatments were detected ($U = 4.5$, $z = -1.433$, $p = 0.190$) [16].

23.3.1.2 Phthalates Detection

PAEs were detected in all the collected soft coral samples (Fig. 23.3a). Water in the aquarium tanks revealed an average concentration of 135 ng/L for the sum of phthalates, with DEHP as the most represented phthalate with an average concentration of 86 ng/L. Soft corals showed an average of total phthalates of 19.2 ng/g. The most represented phthalates were the medium/long chain phthalates DBP, DEHP and short chain phthalate DMP (Fig. 23.3b) [17].

BCF of long chain phthalates resulted as equal to four order of magnitude lower than the predicted BCFs, whereas the short chain phthalates showed experimental BCF from equal to four order of magnitude greater than predicted, with some differences among the surveyed species [17].

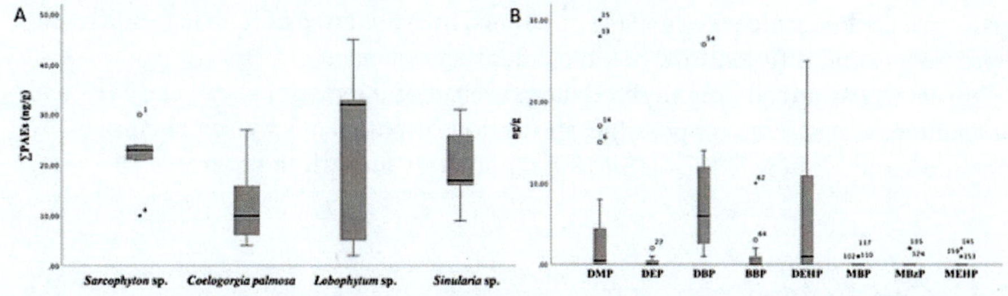

Fig. 23.3 **a** Box plot reporting the phthalates distribution among all the surveyed soft coral species. **b** Box plot reporting the concentration of total phthalates for the different soft coral species. For both the box plots: Line in box = median of sampled concentrations; Box = 25th to 75th percentiles: bars = min and max values excluding outliers, Chemosphere, 297: 134247. Reprinted with permission from Elsevier

23.3.2 Discussion

Plastic debris in the environment contains plasticizers, like phthalates, that can be released during plastic aging or during certain conditions, such as those encountered during the digestion process. Recently, it has been proposed to use the measure of PAEs concentration in marine organisms as an assessment index of their exposure to microplastics. The bioSPME-LC–MS/MS technique may be a useful tool to test such idea, using a common methodology in different environmental situations and allowing to compare same results in completely different situations. This may potentially make possible to assess the efficiency of phthalates as marker of microplastic pollution in diverse marine environments through bioindicators, such as soft corals and sea anemones. Our results at laboratory conditions highlighted that soft corals interact with microplastics at different concentrations both through ingestion and adhesion patterns, as already observed in scleractinian corals [9, 15]. Moreover, all the collected soft coral samples presented PAEs in their tissues. The results indicate that the short chain phthalates DMP and DEP display higher levels of accumulation in the soft coral tissue than expected, while the larger phthalates BBzP and DEHP display lower levels of accumulation. Observation that lowest molecular weight phthalates display BCFs greater than predicted and that higher molecular weight phthalates are below those expected was reported in previous lab and field studies involving other aquatic organisms [18]. This, together with evidence of trophic dilution of the high molecular weight phthalates, is generally considered a proof of metabolic transformation [17]. Since at environmental microplastic concentrations, it is possible to miss or underestimate an organism response resulting from MPs interaction [19], when we set our experimental concentrations, we adopted a higher microplastic concentration range with respect to the environmental one [16].

Therefore, it will be fundamental to verify with the results of the Mediterranean sea anemones samples, the potential of PAEs as marker for the microplastic pollution conditions in seawaters, by comparing the possible microplastic interaction with the organisms, the PAEs presence and the combined interpretation of both in the different environments and different soft benthic cnidarian organisms.

This work has already highlighted the presence of interactions between MPs and sea anemones. Analyses on possible absorption of phthalates by sea anemones by using the same bioSPME-LC–MS/MS methodology are still in progress.

23.4 Conclusions

This study reports that soft corals are able to interact with microplastics through ingestion and adhesion patterns and show their capacity to bioconcentrate metabolized phthalates.

The investigation on *Anemonia viridis* and *Actinia equina* collected on site, will give information not only on the interaction of such organisms with MPs and phthalates at environmental concentrations, but even on the possible use the bioSPME-LC–MS/MS on them. Furthermore, we will investigate the possibility to use them as bioindicators for monitoring the PAEs presence and to explore the potential use of these plasticizers as a marker of microplastic contamination by comparing results coming from different marine environments.

References

1. Teuten EL, Saquing JM, Knappe DRU, Barlaz MA, Jonsson S, Bj¨orn A, Rowland SJ, Thompson RC, Galloway TS, Yamashita R, Ochi D, Watanuki Y, Moore C, Viet PH, Tana TS, Prudente M, Boonyatumanond R, Zakaria MP, Akkhavong K, Ogata Y, Hirai H, Iwasa S, Mizukawa K, Hagino Y, Imamura A, Saha M, Takada H (2009) Transport and release of chemicals from plastics to the environment and to wildlife. Philos Trans R Soc Lond B Biol Sci 364, 2027–2045
2. Gobas FA, Mackintosh CE, Webster G, Ikonomou M, Parkerton TF, Robillard K (2003) Bioaccumulation of phthalate esters in aquatic food-webs. Series Anthrop Comp, pp 201–225
3. Schmidt N, Castro-Jim´enez J, Oursel B, Sempere R (2021) Phthalates and organophosphate esters in surface water, sediments and zooplankton of the NW Mediterranean Sea: exploring links with microplastic abundance and accumulation in the marine food web. Environ Pollut 272, 115970
4. Vered G, Kaplan A, Avisar D, Shenkar N (2019) Using solitary ascidians to assess microplastic and phthalate plasticizers pollution among marine biota: a case study of the Eastern Mediterranean and Red Sea. Mar Pollut Bull 138:618–625
5. Baini M, Martellini T, Cincinelli A, Campani T, Minutoli R, Panti C… Fossi MC (2017) First detection of seven phthalate esters (PAEs) as plastic tracers in superficial neustonic/planktonic samples and cetacean blubber. Anal Methods 9(9), 1512–1520
6. Saliu F, Montano S, Lasagni M, Galli P (2020) Biocompatible solid-phase microextraction coupled to liquid chromatography triple quadrupole mass spectrometry analysis for the determination of phthalates in marine invertebrate. J Chromatogr A 1618:460852
7. Net S, Sempere R, Delmont A, Paluselli A, Ouddane B (2015) Occurrence, fate, behavior and ecotoxicological state of phthalates in different environmental matrices. J Environ Sci Technol 49(7):4019–4035

8. Macali A, Bergami E (2020) Jellyfish as innovative bioindicator for plastic pollution. Ecol Indic 115:106375
9. Hall NM, Berry KLE, Rintoul L, Hoogenboom MO (2015) Microplastic ingestion by scleractinian corals. Mar Biol 162:725–732
10. Steinberg RK, Dafforn KA, Ainsworth T, Johnston EL (2020) Know thy anemone: a review of threats to octocorals and anemones and opportunities for their restoration. Front Mar Sci 7:590
11. Norström AV, Nyström M, Lokrantz J, Folke C (2009) Alternative states on coral reefs: beyond coral–macroalgal phase shifts. Mar Ecol Prog Ser 376:295–306
12. Morais LMS, Sarti F, Chelazzi D, Cincinelli A, Giarrizzo T, Martinelli Filho JE (2020) The sea anemone Bunodosoma cangicum as a potential biomonitor for microplastics contamination on the Brazilian Amazon coast. Environ Pollut 265:114817
13. Jafarabadi AR, Dashtbozorg M, Raudonytė-Svirbutavičienė E, Bakhtiari AR (2021) A potential threat to the coral reef environments: Polybrominated diphenyl ethers and phthalate esters in the corals and their ambient environment (Persian Gulf, Iran). Sci Total Environ 775:145822
14. de Lucia GA, Caliani I, Marra S, Camedda A, Coppa S, Alcaro L, ... Matiddi M (2014) Amount and distribution of neustonic micro-plastic off the western Sardinian coast (Central-Western Mediterranean Sea). Mar Environ Res 100, 10–16
15. Martin C, Corona E, Mahadik GA, Duarte CM (2019) Adhesion to coral surface as a potential sink for marine microplastics. Environ Pollut 255:113281
16. Vencato S, Isa V, Seveso D, Saliu F, Galli P, Lavorano S, Montano S (2021) Soft corals and microplastics interaction: first evidence in the alcyonacean species Coelogorgia palmosa. Aquat Biol 30:133–139
17. Isa V, Saliu F, Bises C, Vencato S, Raguso C, Montano S, Galli P (2022) Phthalates bioconcentration in the soft corals: inter and intra-species differences and ecological aspects. Chemosphere 297:134247
18. Staples CA, Peterson DR, Parkerton TF, Adams WJ (1997) The environmental fate of phthalate esters: a literature review. Chemosphere 35(4):667–749
19. Opitz T, Benítez S, Fernández C, Osores S, Navarro JM, Rodríguez-Romero A, ... Lardies MA (2021) Minimal impact at current environmental concentrations of microplastics on energy balance and physiological rates of the giant mussel *Choromytilus chorus*. Mar Pollut Bull 162:111834

Chapter 24
Random or not? Comparing Microplastic Ingestion and Preys in *Scomber colias* and *Trachurus trachurus*

Laura Ciaralli, Eleonora Monfardini, Tania Pelamatti, Raffaella Piermarini, Cecilia Silvestri, Alice Sbrana, Tommaso Valente, Paolo Tomassetti, and Marco Matiddi

24.1 Introduction

Microplastics represent a rising threat to marine biota [1, 2]. In the last decades, plastic production has grown exponentially, reaching an annual global production of nearly 370 million tons in 2019 [3]. The combination of their chemical-physical properties, such as strength, malleability, corrosion resistance, isolation, and low production cost, has led to the wide use of plastic items, that are eventually discarded and often mismanaged [4]. Hence, the amount of plastic that enters the ocean and marine waters has increased dramatically and special attention is being posed on the impacts of microplastics (MPs) on marine animals.

MPs are plastic particles smaller than 5 mm in size, whose presence has been widely documented in the gastrointestinal tract of several marine species [1, 4]. MPs distribution is uneven and it is influenced by several factors, including distance from potential sources and main currents, that are the major drivers of distribution in the oceans [5]. Given the ubiquity of MPs and the fundamental importance of the fishing

L. Ciaralli (✉) · E. Monfardini · T. Pelamatti · R. Piermarini · C. Silvestri · A. Sbrana · T. Valente · P. Tomassetti · M. Matiddi
ISPRA, Italian National Institute for Environmental Protection and Research, Via Di Castel Romano 100, 00128 Rome, RM, Italy
e-mail: laura.ciaralli.26@gmail.com

T. Valente
Department of Environmental Biology, Univeristy of Rome "La Sapienza", Viale Dell"Università 32, 00185 Rome, RM, Italy

A. Sbrana
Department of Biology, Univeristy of Rome "Tor Vergata", Via Della Ricerca Scientifica 1, 00133 Rome, RM, Italy

M. Cocca et al. (eds.), *Proceedings of the 3rd International Conference on Microplastic Pollution in the Mediterranean Sea*, Springer Water,
https://doi.org/10.1007/978-3-031-34455-8_25

industry as a source of sustenance for human beings [2, 3], microplastic ingestion by fish deserves appropriate attention and, thus, it continues to attract the interest of the scientific community [4]. To address this global issue, it is necessary to understand to what extent microplastics are affecting marine life, also using bioindicator species, in order to plan actions aimed at reducing the amount of microlitter that reaches the marine environment [6, 8]. Despite previous studies suggesting that the feeding behavior of different fish species may affect the occurrence and the diversity of ingested microplastic types [4], little is known about the selection mechanisms that determine the pathways of microplastics through the food webs.

For this purpose, this study aims at shedding light on the mechanisms that influence microplastic ingestion in two commercial fish species both displaying a pelagic feeding behavior [6], namely the Atlantic chub mackerel *Scomber colias*, and the Atlantic horse mackerel *Trachurus trachurus*, from Anzio Coast waters, Tyrrhenian Sea, Western Mediterranean.

S. colias geographic distribution comprehends the North Atlantic Ocean and Mediterranean Sea, from 0 to 200 m in depth, whereas *T. trachurus* ranges from Norway to South Africa, including Mediterranean Sea, and it is more common from 100 to 200 m in depth [5, 6].

24.2 Experimental

Fish were caught using gillnets set in front of the Anzio Coast (Tyrrhenian Sea, Western Mediterranean) in spring 2021. A total of 124 samples (55 *S. colias* and 69 T. *trachurus*) were collected at landing and immediately stored at −20 °C until laboratory analysis.

A subsample of 25 *S. colias* (mean ± sd total length, TL = 30.3 ± 2.5 cm) and 25 T. *trachurus* (TL = 25.6 ± 2.2 cm) was used for microplastic analysis following the guidelines provided by Matiddi et al. [9]. Briefly, gastrointestinal tracts underwent alkaline digestion (10% KOH) to degrade the organic matter. Then, solutions were filtered on glass microfiber membranes (Whatman GF/D™; pore size: 2.7 μm) and subsequently observed under a dissecting microscope. Microplastics were characterized by shape and the polymer composition of all items was confirmed using a Nicolet iN5 Fourier Transform Infrared Micro-Spectrometry (μFT-IR, Thermo Fisher Scientific, Madison, WI, USA). The remaining 30 *S. colias* (TL = 29.3 ± 2.6 cm) and 44 T. *trachurus* (TL = 24.4 ± 1.9 cm) were used for prey identification: stomachs were dissected, and preys were identified to the lowest possible taxonomic level, counted under a dissecting microscope, and weighed with a precision of 0.01 g.

24.2.1 Quality Assurance and Quality Control

All the precautions suggested by Matiddi et al. [9], were adopted during every analytical step to reduce secondary contamination. One blank control was processed together with every batch of 5 samples in order to check for possible contamination. Since blank controls only revealed the presence of microfibers made of natural polymers (cellulosic-made or wool), no results adjustment techniques were adopted for presenting data on synthetic fibers and the other MP types.

24.2.2 Data and Statistical Analysis

Frequency of occurrence (FO %) was computed as follow:

$$FO\% = (\text{No. of individuals with ingested MPs} * \text{No. of individuals examined}^{-1}) * 100$$

Differences in FO were tested using Generalized Linear Models (GLMs) assuming a binomial error structure and logit link function. The significance level was set at $p < 0.05$.

24.3 Results and Discussion

MP ingestion was detected with an overall frequency of occurrence of 46.0% and no significant differences between the two examined species (48.0% in *S. colias* and 44.0% in *T. trachurus*; p-value > 0.05). The 23 individuals with ingested MPs contained a total of 33 items (mean ± sd = 0.66 ± 0.93): 17 synthetic fibers, 12 fragments, 2 foams, 1 film and 1 filament. Pie charts representing the diversity of MP types ingested by the two species are available in Fig. 24.1. Excluding fibers and filaments (FO = 32.0% and 28.0% in *T. trachurus* and *S. colias*, respectively), statistical analyses revealed a significant higher frequency of the other MP types in *S. colias* (FO = 32.0%) than *T. trachurus* (FO = 8.0%) (p-value < 0.05).

Stomach content analysis for prey identification revealed that *S. colias* in our sample fed mostly on Thaliaceans and Crustacea, with a small percentage of Mollusca, as shown in Fig. 24.2. Among crustaceans, *Chlorotocus crassicornis* exhibited the most relevant contribution (Table 24.1). Differently, the *T. trachurus* samples showed a diet based more on Crustacea, Teleostei, and Mollusca, with a prevalence of Mysidia. Detailed data on the diet composition of the two species are available in Fig. 24.2 and Table 24.1.

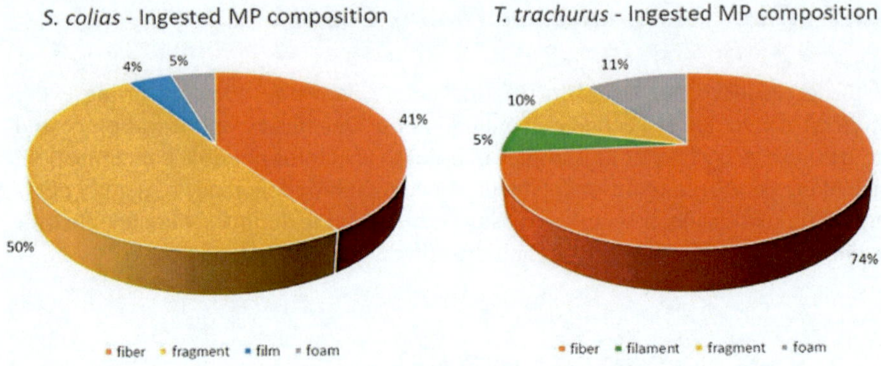

Fig. 24.1 Pie charts for microplastic categories ingested by *Scomber colias* and *Trachurus trachurus* sampled in 2021 out off Anzio coast (Tyrrhenian Sea, Western Mediterranean)

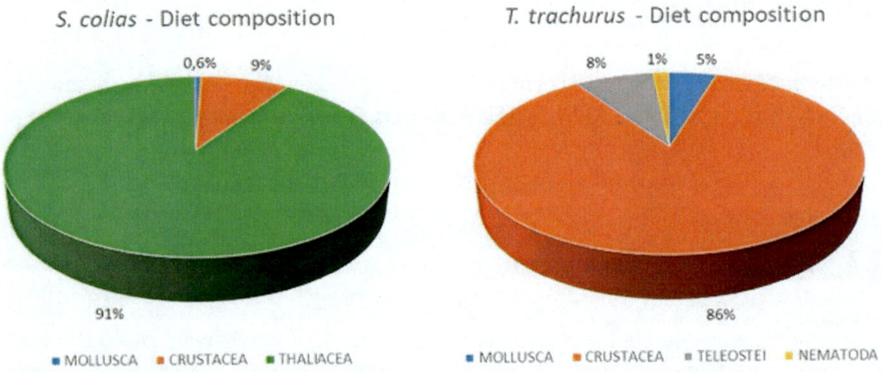

Fig. 24.2 Pie charts for diet composition of *S. colias* and *T. trachurus*. Sampled in 2021 out off Anzio coast (Tyrrhenian Sea, Western Mediterranean)

The integration of data on MP ingestion and diet composition provided some interesting suggestions. Indeed, the benthopelagic diet of *T. trachurus* and the pelagic-feeder behavior of *S. colias* resulted consistent with literature data [6, 7]. Our preliminary data seemed to suggest that *T. trachurus* makes an active selection of fast-moving preys (mostly mysids and teleosts) and ingests almost only threadlike microplastics, such as fibers and filaments. These types of MPs tend to sink due to biofouling and water mixing and, therefore, they are expected to be found in the bentho-pelagic zone. In contrast, the high frequency of thaliaceans in *S. colias* revealed a more planktivorous behavior, which is associated to the ingestion of a wider variety of microplastic types (including also films, fragments and foams). Foams, fragments, and films tend to float closer to the surface, and even after sinking, they have the tendency to refloat because of their bigger volume (and consequently higher rising velocity) compared to fibers and filaments [10]. Therefore, those floating MPs are predictable to be found in a pelagic feeder, like *S. colias*. In this view, our results

Table 24.1 List of the species found in the stomach contents of *S. colias* (30 individuals) and *T. trachurus* (44 individuals) sampled in 2021 off Anzio coast (Tyrrhenian Sea, Western Mediterranean)

Prey	*S. colias*	*T. trachurus*
Mollusca		
Cephalopoda ind	2	1
Gastropoda ind	–	1
Turritellinella tricarinata	–	1
Crustacea		
Chlorotocus crassicornis	18	–
Crustacea ind	2	3
Decapoda ind	8	–
Mysida ind	–	52
Penaeideae ind	1	–
Teleostei		
Teleostei ind	–	5
Nematoda		
Nematoda ind	–	1
Thaliacea		
Thaliacea ind	~300	–

seemed to suggest that the microplastic ingestion patterns observed in the two examined species could be linked to the existing differences between their prey searching modalities and feeding depth.

24.4 Conclusions

The wide distribution of plastic makes interaction with marine biota likely to happen, leading to harmful consequences on marine life [10]. Microplastic ingestion has been widely reported in several fish species [8, 11], and previous studies suggested that the occurrence and the diversity of ingested microplastic types may be affected by the feeding behavior of different fish species [1]. In this study we combined microplastic analysis and diet composition analysis of two pelagic species, the Atlantic chub mackerel *Scomber colias* and the Atlantic horse mackerel *Trachurus trachurus*, aiming at clarifying the possible correlation between their feeding habits and microplastic ingestion.

According to our results, the microplastic ingestion patterns observed in the two species do not occur randomly, rather it could be related to existing differences between prey searching modalities and feeding depth, which make *S. colias* and *T. trachurus* exposed to different types and amount of MPs. Moreover, microplastic pollution is a constantly evolving research topic, as well as an urgent issue, that requires future studies to better profile the interaction dynamics between MPs pollution and marine organisms.

References

1. Valente T, Scacco U, Matiddi M (2020) Macro-litter ingestion in deep-water habitats: is an underestimation occurring? Environ Res 186:109556
2. Galloway TS, Cole M, Lewis C (2017) Interactions of microplastic debris throughout the marine ecosystem. Nat Ecol Evol 1(5)
3. FAO-FIGIS (2005) A world overview of species of interest to fisheries. Chapter: Trachurus trachurus. Retrieved on 05 May 2005, from www.fao.org/figis/servlet/species?fid=2306.3p. FIGIS Species Fact Sheets. Species Identification and Data Programme-SIDP, FAO-FIGIS
4. Bergmann M, Gutow L, Klages M (2015) Marine anthropogenic litter. Springer Nature
5. Sánchez-Hernández LJ, Ramírez-Romero P, Rodríguez-González F, Ramos-Sánchez VH, Montes RAM, Rubio HR, Jonathan (2021) Seasonal evidences of microplastics in environmental matrices of a tourist dominated urban estuary in Gulf of Mexico, Mexico. Chemosphere 277:130261
6. C JRC, European Commission, Joint Research Centre (2013) MSFD technical subgroup on marine litter (TSG-ML). Guidance on Monitoring of Marine Litter in European Seas
7. Dhimmer VR (2017) Microplastics in gastrointestinal tracts of *Trachurus trachurus* and *Scomber colias* from the Portuguese Coastal waters. Doctoral dissertation
8. GESAMP (2019) Guidelines for the monitoring and assessment of plastic litter and microplastics in the ocean. In: Kershaw PJ, Turra A, Galgani F (eds) (IMO/FAO/UNESOIOC/UNIDO/WMO/IAEA/UN/UNEP/UNDP/ISA joint group of experts on the scientific aspects of marine environmental protection). Rep. Stud, vol 99. GESAMP No., pp 8–12
9. Matiddi M, Pham CK, Anastasopoulou A, Andresmaa E, Avio CG, Bianchi J, Chaieb O, Palazzo L, Darmon G, de Lucia GA, Deudero S, Sozbilen D, Eriksson J, Fischer E, Gómez M, Herrera A, Hattia E, Kaberi H, Kaska Y, Kühn S, Lips I, Miaud C, Gambaiani (2021) Monitoring micro-litter ingestion in marine fish: a harmonized protocol for MSFD and RSCs areas. INDICIT II Project. https://indicit-europa.eu/cms/wp-content/uploads/2021/06/Monitoringmicrolitter-ingestion-in-marine-fish-1.pdf
10. Klink D, Peytavin A, Lebreton L (2022) Size dependent transport of floating plastics modeled in the global ocean. Front Mar Sci 9:1–9. https://doi.org/10.5281/zenodo.6301282
11. Clara L et al (2020) Microplastic ingestion and diet composition of planktivorous fish. Limnol Oceanogr Lett 5(1):103–112

Chapter 25
May Mesopelagic Fishes Play an Important Role as Vector of Microplastics Across the Mediterranean Trophic Web? A Case of Study in the Strait of Messina

Federica Laface, Cristina Pedà, Matteo Nannini, Giuseppe Cangemi, Valentina Sciutteri, Pietro Battaglia, and Teresa Romeo

25.1 Introduction

Currently, the issue of Microplastics (MPs) pollution has become of global concern. MPs are defined as particles < 5 mm [1] and can be manufactured as industrial microbeads and scrubbers (primary MPs) or originate from the fragmentation of larger plastics (secondary MPs) free in the marine environment resulting from abiotic and/or biotic factors [2]. MPs, ubiquitous in marine environment worldwide, can enter the trophic web through direct or secondary ingestion by marine fauna [3]. MPs ingestion represents a serious threat to marine organisms, causing physical/mechanical damage and chemical damage [4].

Although many studies have documented the MPs ingestion [4], the research challenge focused now on assessing the transfer of MPs across marine trophic levels.

F. Laface (✉) · C. Pedà · M. Nannini · G. Cangemi · V. Sciutteri · P. Battaglia · T. Romeo
Department of Integrative Marine Ecology (EMI), Stazione Zoologica Anton Dohrn - National Institute of Biology, Ecology and Marine Biotechnology, Sicily Marine Centre, Villa Pace - Contrada Porticatello 29, 98167 Messina, Italy
e-mail: federica.laface@szn.it

F. Laface
Department of Chemical, Biological, Pharmaceutical and Environmental Sciences, University of Messina, Viale Ferdinando Stagno D'Alcontres 31, 98166 Messina, Italy

T. Romeo
Institute for Environmental Protection and Research, ISPRA, Via Dei Mille 56, 98057 Milazzo, ME, Italy

M. Cocca et al. (eds.), *Proceedings of the 3rd International Conference on Microplastic Pollution in the Mediterranean Sea*, Springer Water,
https://doi.org/10.1007/978-3-031-34455-8_26

MPs ingestion has also been reported in deep environment and specifically in mesopelagic fish from several oceanic and Mediterranean areas [5–12].

However, the role of mesopelagic fish in the MPs' transfer across the marine trophic web is still poorly explored. They play an important ecological role in the food web, performing diel vertical migrations and contributing to the connection between epipelagic waters to deep-sea ecosystems [12, 13]. Indeed, they are involved in the energy transfer from zooplankton to higher trophic levels, representing an important food source for medium and large pelagic predators [14, 15].

In this view, the present paper investigates: (i) the MPs ingestion in five mesopelagic fish belonging to the families Myctophidae (*Electrona risso, Hygophum benoiti, Myctophum punctatum*) and Sternoptychidae (*Argyropelecus hemigymnus* and *Maurolicus muelleri*) from the Strait of Messina; (ii) the potential transfer of MPs to intermediate and large predators, across the pelagic trophic web, mediated by mesopelagic food resources.

25.2 Experimental

An integrated approach based on the study of both MPs ingestion by mesopelagic species and their trophic relationships in the study area was applied to assess MPs transfer in the trophic web.

25.2.1 Sample Collection

Sample collection was carried out along the shore of the Sicilian coast of the Strait of Messina (central Mediterranean Sea) in 2020–2021. This area is characterized by peculiar hydrodynamic system, which periodically allows the stranding of deep-sea fauna [16]. Three Myctophidae (*E. risso, H. benoiti* and *M. punctatum*) and two Sternoptychidae (*A. hemigymnus* and *M. muelleri*) were selected for this study. For each species, the gastrointestinal tracts (GITs) were removed and grouped into size classes (*E. risso*: <20 mm, 20–40 mm, >40 mm; *H. benoiti*: <25 mm, 25–40 mm, >40 mm; *M. punctatum*: 30–50 mm, >50 mm; *A. hemigymnus*: <20 mm, 20–30 mm, >30 mm; *M. muelleri*: <20 mm; 20–40 mm; >40 mm). The GITs were weighed and placed into glass beakers for the analysis.

25.2.2 *Isolating Microplastics*

25.2.2.1 Chemical Digestion

MPs were extracted from GITs using the chemical digestion protocol of Schirinzi et al. [17], which involves two digestion steps. During the first step, each size group was treated with a potassium hydroxide solution (10% KOH, ratio 1:3 (w/v)) and incubated in a stove at 60 °C/6 h. The digested solutions were vacuum filtered through a glass fibre membrane (pore size 1.6 μm). During a second digestion step, a 20% HNO_3 solution was added on the clogged filters for 60 min at room temperature to remove most of the organic and inorganic material.

25.2.2.2 Polymer Identification

Filters were examined under a stereomicroscope Zeiss Discovery V.8. coupled with Axiocam 208 camera. All isolated particles were counted, measured (length and width in mm) and photographed. Polymer nature was identified by Fourier transform infrared (FT-IR) spectroscopy technique using the Agilent Cary 630 spectrometer supplied with specific polymer libraries and the level of certainty to match the sample spectrum with reference spectra was set up to >70% [17, 18]. The average number of plastic items found in the GITs was calculated on the total number of individuals (N. plastic items/N. all examined individuals).

25.2.2.3 Microplastics Transfer Across the Trophic Web

To assess the potential transfer of MPs from mesopelagic fish to pelagic predators, we collected the available information on the feeding habits of intermediate and top predators from the study area. We selected the studies where mesopelagic species represent an important food resource for predators of the Strait of Messina [14, 15, 19] and also used unpublished data to improve information on trophic relationship between pelagic resources. Overall, we considered the top predators *Thunnus thynnus* and *Xiphias gladius* as well as the intermediate predators *Chauliodus sloani* (Stomiidae), *Trachurus picturatus* (Carangidae), *Lepidopus caudatus* (Trichiuridae) and *Ommastrephid squids* (Ommastrephidae).

Since the values of prey abundance of mesopelagic fish in stomach of these predators, we hypothesized a MPs transfer scheme across the trophic web.

25.3 Results and Discussion

A total of 1101 GITs of mesopelagic fish were examined. Table 25.1 shows the mean size (SL mm) per size group of each species.

Overall, 41 plastic particles (0.04 items/specimens) were detected in mesopelagic GITs (Table 25.1), mainly belonging to small MPs (61%) and large MPs (34%), although mesoplastics (5%) were also found.

The most abundant polymers identified by FT-IR analysis were polyvinylchloride (PVC; 12%), nylon (polyamide, PA; 10%), polytetrafluoroethylene (PTFE; 10%) and polyethylene (PE; 7%). Other polymers were polypropylene (PP; 5%), rubber (5%), polystyrene (PS; 2%) and polyurethane (PUR; 2%). The highest ratio between the ingested MPs items and the number of examined individuals was found in *H. benoiti* (0.06 plastic items/individual). This agrees with the findings of Romeo et al. [12], although this last study was only based on visual inspection of stomach contents of

Table 25.1 Number of individuals and mean size (SL mm) per size group of each mesopelagic species investigated. Abundance of plastic ingestion by each size class pool of mesopelagic species from the Strait of Messina

Family/species	Size group (SL mm)	N. specimens	Mean SL ± Standard deviation	N. MPs	N plastic items/total examined individuals
MYCTOPHIDAE					
Electrona risso	<20 mm	180	15.6 ± 2.8	3	0.02
	20–40 mm	63	28.4 ± 6.2	6	0.10
	>40 mm	10	43.1 ± 2.4	0	0.00
	Total	253	19.9 ± 8.2	9	0.04
Hygophum benoiti	<25 mm	107	23.0 ± 1.7	2	0.02
	25–40 mm	128	33.2 ± 3.5	10	0.08
	>40 mm	104	45.3 ± 3.0	9	0.09
	Total	339	33.7 ± 9.3	21	0.06
Myctophum punctatum	30–50 mm	26	44.6 ± 4.3	2	0.08
	>50 mm	96	58.4 ± 3.2	2	0.02
	Total	122	55.5 ± 6.6	4	0.03
STERNOPTYCHIDAE					
Argyropelecus hemigymnus	<20 mm	105	15.8 ± 2.9	0	0.00
	20–30 mm	117	27.6 ± 2.2	4	0.03
	>30 mm	107	29.5 ± 3.4	3	0.03
	Total	329	25.0 ± 7.0	7	0.02
Maurolicus muelleri	20–40 mm	36	27.5 ± 4.1	0	0.00
	>40 mm	22	44.6 ± 2.3	0	0.00
	Total	58	33.9 ± 9.1	0	0.00
Total		1101	30.3 ± 13.2	41	0.04

lanternfish. In addition, the greatest polymeric variability of MPs was also observed in *H. benoiti*. The variation in ingestion rates of MPs by investigated species maybe related to their different feeding and migrating behavior. Moreover, these species hunt using vision, so they may mistake plastics for prey by intentionally ingesting MPs [10, 12].

The assessment of MPs transfers throughout the trophic web of the Strait of Messina, mediated by mesopelagic food resources, is summarized in Fig. 25.1 that defined the trophic relationships between mesopelagic fish and their predators in the study area. Bluefin tuna directly prey upon mesopelagic fishes in the Strait of Messina, exploiting upwelling phenomena to forage on myctophids and sternoptychids, but also feeds on important intermediate predators such as *C. sloani*, *T. picturatus* and ommastrephid squids. Therefore, it can partially collect MPs through direct ingestion of mesopelagic fish. Instead, swordfish may only indirectly accumulate MPs of mesopelagic fish origin by preying upon ommastrephid squids and *L. caudatus*. In Fig. 25.1, the green lines indicate the trophic relationships between mesopelagic species and intermediate predators, while the red lines the relationships between top predators and both intermediate predators and mesopelagic organisms. Although the ingestion rate of MPs by mesopelagic fish was low, the analysis of the complex trophic relationships in the study area and the amounts of mesopelagic specimens ingested by predators suggest that the concentration of MPs transferred across trophic levels can reach important orders of magnitude at the higher levels.

25.4 Conclusions

The present research provides preliminary results on the potential transfer of MPs in trophic web of the Strait of Messina mediated by mesopelagic fish resources and underlines the importance of these species in the MPs transfer across the food web. Further studies are needed to better understand how long MPs stay in the GITs and how quickly they are evacuated. An important perspective will be to assess the implications for human consumption related to the potential effects of contaminants the transfer from litter to edible fish tissue.

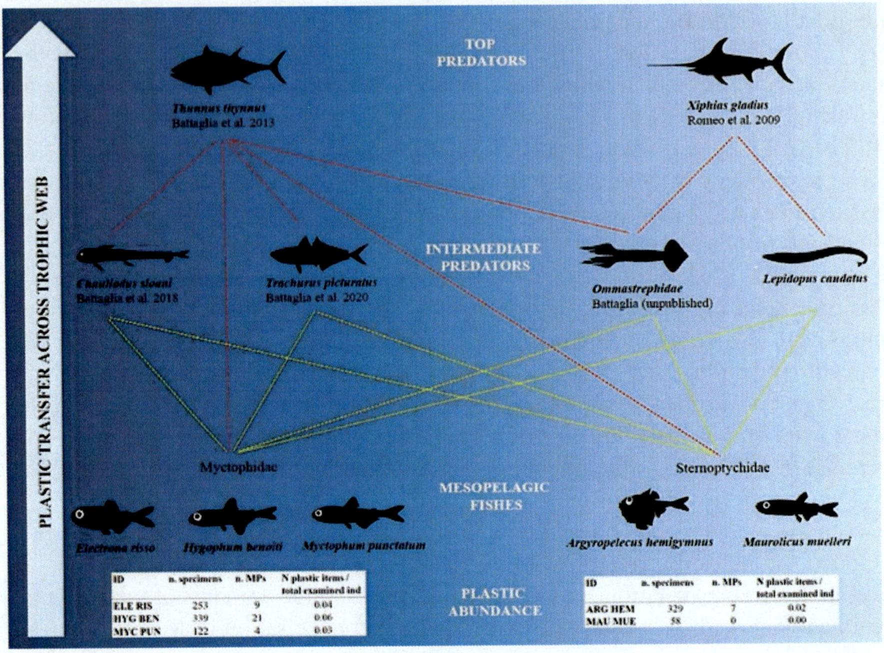

Fig. 25.1 Microplastics transfer across the trophic web in the Strait of Messina mediated by mesopelagic food resources

References

1. Arthur C, Baker JE, Bamford HA (2009) Proceedings of the international research workshop on the occurrence, effects, and fate of microplastic marine debris, September 9–11, 2008. University of Washington Tacoma, Tacoma, WA, USA
2. Andrady AL (2011) Microplastics in the marine environment. Mar Pollut Bull 62:1596–1605
3. Pedà C, Battaglia P, D'Alessandro M, Laface F, Malara D, Consoli P, Vicchio TM, Longo F, Andaloro F, Baini M, Galli M, Bottari T, Fossi MC, Greco S, Romeo T (2020) Coupling gastro-intestinal tract analysis with an airborne contamination control method to estimate litter ingestion in demersal elasmobranchs. Front Environ Sci 8:119
4. Fossi MC, Pedà C, Compa M, Tsangaris C, Alomar C, Claro F, Ioakeimidis C, Galgani F, Hema T, Deudero S, Romeo T, Battaglia P, Andaloro F, Caliani I, Casini S, Panti C, Baini M (2018) Bioindicators for monitoring marine litter ingestion and its impacts on Mediterranean biodiversity. Environ Pollut 237:1023–1040
5. Boerger CM, Lattin GL, Moore SL, Moore CJ (2010) Plastic ingestion by planktivorous fishes in the North Pacific Central Gyre. Mar Pollut Bull 60:2275–2278
6. Davison P, Asch RG (2011) Plastic ingestion by mesopelagic fishes in the North Pacific Subtropical Gyre. Mar Ecol Prog Ser 432:173–180
7. Lusher AL, O'Donnell C, Officer R, O'Connor I (2016) Microplastic interactions with North Atlantic mesopelagic fish. ICES Mar Sci Symp 73:1214–1225
8. Wieczorek AM, Morrison L, Croot PL, Allcock AL, MacLoughlin E, Savard O, Brownlow H, Doyle TK (2018) Frequency of microplastics in mesopelagic fishes from the Northwest Atlantic. Front Mar Sci 5

9. McGoran AR, Maclaine JS, Clark PF, Morritt D (2021) Synthetic and semi-synthetic microplastic ingestion by mesopelagic fishes from Tristan da Cunha and St. Helena, South Atlantic. Front Mar Sci 8
10. Justino AKS, Ferreira GVB, Schmidt N, Eduardo LN, Fauvelle V, Lenoble V, Sempéré R, Panagiotopoulos C, Mincarone MM, Frédou T, Lucena-Frédou F (2022) The role of mesopelagic fishes as microplastics vectors across the deep-sea layers from the Southwestern Tropical Atlantic. Environ Pollut 300:118988
11. Zhu L, Wang H, Chen B, Sun X, Qu K, Xia B (2019) Microplastic ingestion in deep-sea fish from the South China Sea. Sci Total Environ 677:493–501
12. Romeo T, Pedà C, Fossi MC, Andaloro F, Battaglia P (2016) First record of plastic debris in the stomach of Mediterranean lanternfishes. Acta Adriatica Int J Mar Sci 57:115–122
13. Gjøsaeter J, Kawaguchi K, Nations F, A.O. of the U. (1980) A review of the world resources of mesopelagic fish, food & agriculture org
14. Battaglia P, Andaloro F, Consoli P, Esposito V, Malara D, Musolino S, Pedà C, Romeo T (2013) Feeding habits of the Atlantic bluefin tuna, *Thunnus thynnus* (L. 1758), in the central Mediterranean Sea (Strait of Messina). Helgol Mar Res 67:97–107
15. Battaglia P, Pagano L, Consoli P, Esposito V, Granata A, Guglielmo L, Pedá C, Romeo T, Zagami G, Vicchio TM, Guglielmo R, Andaloro F (2020) Consumption of mesopelagic prey in the Strait of Messina, an upwelling area of the central Mediterranean Sea: feeding behaviour of the blue jack mackerel *Trachurus picturatus* (Bowdich, 1825). Deep-Sea Res I: Oceanogr Res Pap 155:103158
16. Battaglia P, Ammendolia G, Cavallaro M, Consoli P, Esposito V, Malara D, Rao I, Romeo T, Andaloro F (2017) Baseline data to characterize and manage the small-scale fishery (SSF) of an oncoming Marine Protected Area (Cape Milazzo, Italy) in the western Mediterranean Sea. Ocean Coast Manag 148:231–244
17. Schirinzi GF, Pedà C, Battaglia P, Laface F, Galli M, Baini M, Consoli P, Scotti G, Esposito V, Faggio C, Farré M, Barceló D, Fossi MC, Andaloro F, Romeo T (2020) A new digestion approach for the extraction of microplastics from gastrointestinal tracts (GITs) of the common dolphinfish (*Coryphaena hippurus*) from the western Mediterranean Sea. J Hazard Mater 397:122794
18. Pedà C, Longo F, Berti C, Laface F, De Domenico F, Consoli P, Battaglia P, Greco S, Romeo T (2022) The waste collector: information from a pilot study on the interaction between the common octopus (*Octopus vulgaris*, Cuvier, 1797) and marine litter in bottom traps fishing and first evidence of plastic ingestion. Mar Pollut Bull 174:113185
19. Battaglia P, Ammendolia G, Esposito V, Romeo T, Andaloro F (2018) Few but relatively large prey: trophic ecology of *Chauliodus sloani* (Pisces: Stomiidae) in deep waters of the Central Mediterranean Sea. J Ichthyol 58:8–16

Chapter 26
Microplastic Fiber Content in *Mullus barbatus* from the Tyrrhenian Sea: Preliminary Results

Serena Santonicola, Michela Volgare, Emilia Di Pace, Mariacristina Cocca, Gennaro Raimo, and Giampaolo Colavita

26.1 Introduction

Marine litter ingestion in Mediterranean fish species has gained more interest in the last few years, due to the possible link with their commercial and economic value as fishery resources [1, 2]. The Mediterranean Sea is a semi-enclosed, highly populated basin, exposed to heavy coastal pressures that increase marine debris including marine plastics [3, 4]. In this context, new emerging debris have been considered, represented by synthetic of petrochemical origin microfibers (i.e. polyester, polyamide, polypropylene, etc.) and natural fibers (i.e. cotton, wool), coming from the textile industry or urban centers wastewaters [1, 5]. Microfiber ingestion was reported both for pelagic and demersal fish species from the Mediterranean Sea [1, 6–8]. Benthic fish may be more exposed to fibers than pelagic fish due to the fact that one of the most common types of fibers consist of nylon, mostly used in fishing gear, which has negative or neutral buoyancy and settles in the sediment [9, 10].

S. Santonicola (✉) · G. Raimo · G. Colavita
Department of Medicine and Health Sciences "V. Tiberio", University of Molise, Via F. De Santis, 86100 Campobasso, Italy
e-mail: serena.santonicola@unimol.it

S. Santonicola · M. Volgare · E. Di Pace · M. Cocca
Institute of Polymers, Composites and Biomaterials National Research Council of Italy, Via Campi Flegrei, 34 80078 Pozzuoli (NA), Italy

M. Volgare
Department of Chemical Engineering, Materials, and Industrial Production, University of Naples Federico II, Naples, Italy

M. Cocca et al. (eds.), *Proceedings of the 3rd International Conference on Microplastic Pollution in the Mediterranean Sea*, Springer Water,
https://doi.org/10.1007/978-3-031-34455-8_27

The occurrence of fibrous microplastics and natural microfibers in commercially important fish species may pose a potential risk to human health [11, 12], considering that microplastics were found also in the edible tissues of fish species that are commonly used for human consumption [13, 14].

In the light of this information, aim of the study was to assess microfiber ingestion in *Mullus barbatus* (Linnaeus, 1758), commonly called the red mullet, a commercial benthic fish species widely spread in the Mediterranean Sea, where it inhabits sandy and muddy bottoms. Red mullets swallow sediment together with the prey, after identifying them with their barbels, and expelling then the sediment through their gills. Considering their feeding behaviour, that may increase the risk of accidently ingesting plastic, and the reduced mobility, they have been proposed as a sentinel species for different pollutants, including microplastics [1, 3, 10, 15].

26.2 Experimental

26.2.1 Materials

Sodium Chloride, Hydrogen Peroxide solution 30%, and Potassium hydroxide were purchased from Carlo Erba (Val De Reuil, France). Cellulose nitrate (pore size 8 μm) and cellulose acetate (pore size 0.45 μm) filters were provided by Sartorius Stedim Biotech (Gottingen, Germany). The filtrating system was supplied by Advantec (Dublin, CA 94,568, USA).

26.2.2 Methods

Red mullets (n = 20) from the Tyrrhenian Sea (Western Mediterranean Sea), collected at fish markets, were transported to the laboratories in coolers with ice packs, and then stored in a freezer at −20 °C until analyses. Fish were thawed, washed with filtered distilled water, measured, and weighted (Table 26.1).

Then, tissue samples were prepared in a previously cleaned laboratory with limited access to prevent microplastic contamination. For the correction of potential procedural contamination, one blank control without any tissue was carried out for every sample group (5–6 individuals) processed on the same day. Gastrointestinal samples were dissected and transferred individually into a glass Erlenmeyer flask

Table 26.1 Mean weight and length of *Mullus barbatus* samples

	Mean weight (g) ± SD	Mean length (cm) ± SD
Red mullet (N.20)	58.5 ± 29.6	13.78 ± 2.35

and submerged with a 10% KOH solution, approximately triple the volume of the tissue. The samples were incubated overnight at 45 °C to digest the organic material. Density separation was performed to isolate microplastics by the use of saturated sodium chloride solution (1.2 g/mL NaCl). The overlying water was vacuum filtered using cellulose nitrate filters (pore size 8 μm), then the filters with retained materials were placed into clean Petri dishes, and to adequately digest all tissue residues, a 15% H_2O_2 solution was added to the membranes and allowed to dry in oven (45 °C, overnight), before the microscopical observation.

During this first phase, only the microfibers on the filter were identified, and counted. The fibers were carefully observed under a light microscope (LEICA M205C, magnification of 0.78–16×), and classified according to their morphological features: cross-section shape, patterns, breakages and alterations of the fiber body, shape and appearance of the ends, according to Rodríguez-Romeu et al. [3] and Santonicola et al. [16] (Figs. 26.1, 26.2).

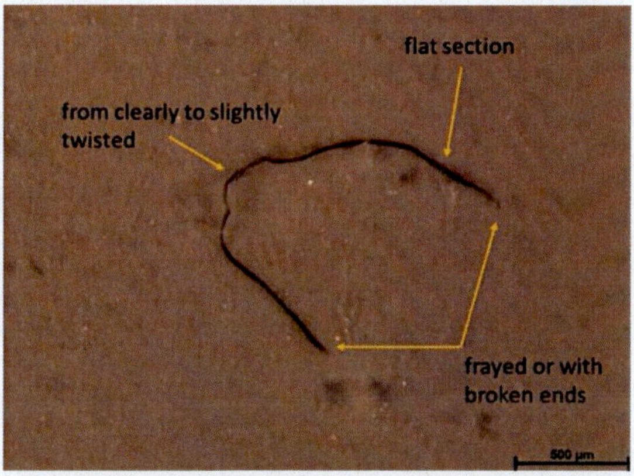

Fig. 26.1 Optical micrograph of a natural fiber recovered from filter surface: description of some morphological features

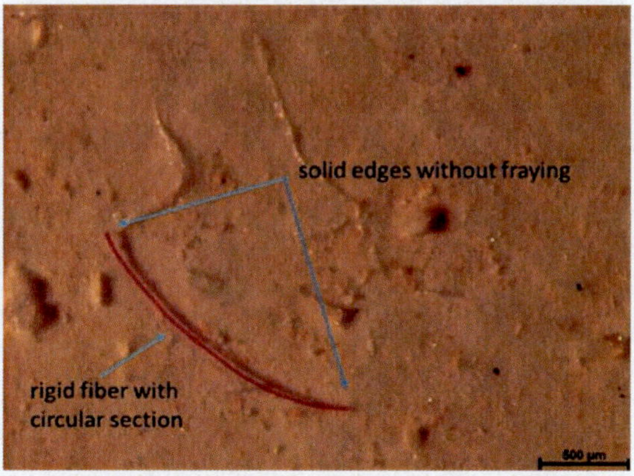

Fig. 26.2 Optical micrograph of a synthetic fiber recovered from filter surface: description of some morphological features

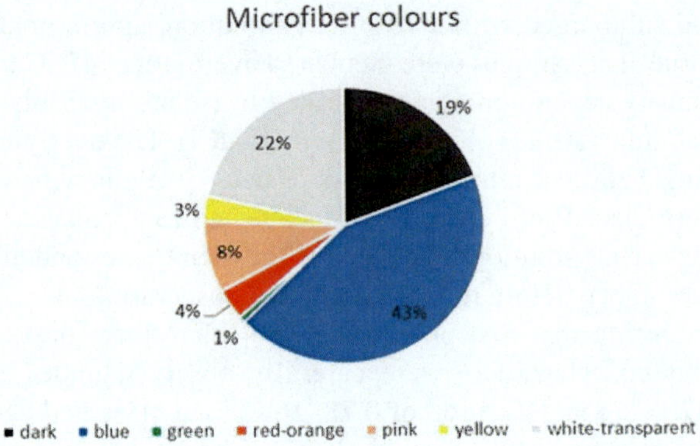

Fig. 26.3 Percentage distribution of microfibers colours in red mullets

26.3 Results and Discussion

26.3.1 Results

A total of 274 microfibres were recorded. The procedural blanks contained 8.28 (mean value) microfibers/filter. The percentage of fish with microfibres was 55% of which 90% ingested more than one microfiber. On average, red mullet contained 7.44 microfibers/individual. Visual characterization of fibers by typologies allowed classification of 65% of the items as natural microfibres. The most common colours were blue (43%), transparent (22%), and black (19%) (Fig. 26.3).

26.3.2 Discussion

Seafloor habitats are considered the ultimate sink for plastics in the marine environment [10]. Thus, higher loads of microplastics would be expected in sediments than in the water column and consequently in species feeding on the seafloor. A study conducted on the different seabed located in the Tyrrhenian Sea confirmed high quantity of plastic debris on seafloor [4]. Wastewater from washing machines is considered the main source of microfibers in sediments, taking to account the proximity of the Tyrrhenian Sea to populated areas [5].

Microfibers are the major abundant shape found in *M. barbatus* from different Mediterranean areas [1, 6, 8]. Half of the red mullets analysed in the present study presented microfibers (synthetic and natural fibres) in their digestive tract. *M. barbatus* that ingested microplastics vary between different studies, as reported for the Turkish (42%), Greek (32%) and Spanish (19%) Mediterranean coasts and for Adriatic Sea (64%) [1, 6, 8, 10, 15]. These differences may also depend on the methodology used in each case. Lower values are found when comparing microplastics (1.75 items/ind.) found by Bellas et al. [10] versus the average number of microfibers per individual in our study. This comparison, however, must be considered with caution because the authors do not report on non-synthetic fibres [3].

Different studies, in fact, did not include natural fibers despite they may represent a consistent fraction of total fibers in some Mediterranean marine environments [3, 7]. According to Rodríguez-Romeu et al. [3], our results show that 65% of the fibres in the gastrointestinal tract of *M. barbatus* are natural. The faster degradation of natural fiber compared to the synthetic one makes it easier to release toxic compounds and organic pollutant adsorbed to the surface. The potential risks of transfer these chemicals to the edible parts of fish are of great concern for human health and may have serious economic consequences for fishing activity [9].

As regarding colours of microfibers, our findings were similar to other studies [4, 10, 17]. The occurrence of microfibers with intense blue colour could be related to their abundance in sea water [18]. Transparent microfibers may be ingested because similar to prey, as well as young specimens could be more exposed to black fibers due to the fact that are similar to their food. However, the mean fish length (Table 26.1) suggested that the analysed samples of red mullet are adult specimens [19]. Therefore, the occurrence of numerous black fibers, confirmed as in this species the accidental microplastics ingestion, when fish detecting preys with their barbells, is an important source of exposure [4, 10, 11].

26.4 Conclusion

Human activities play an important role in the microplastic distribution in marine environments. The high occurrence of microfibers in marine species might reflect a wide distribution of textile fibers along densely populated coastlines. Preliminary results confirmed the microfiber ingestion by red mullet from the Tyrrhenian Sea, Western Mediterranean Sea.

Recently, microplastics have been found also in the edible tissues of fish species that are commonly used for human consumption, but little is known about microfibers. This risk needs to be deeply assessed through further investigation on commercial fish species, considering both natural and synthetic microfibers and the adsorbed contaminants. In this context, the adopted visual approach may be useful to differentiate synthetic and natural fibers, representing a fast and easy method to gain information about the fibrous microplastics in complex matrices.

Acknowledgements This study was supported by the research project "SIstemi di Rilevamento dell'Inquinamento MArino da Plastiche e successivo recupero-riciclo. SIRIMAP PON_project" from the Ministry of Education, Italy.

References

1. Giani D, Baini M, Galli M, Casini S, Fossi MC (2019) Microplastics occurrence in edible fish species (*Mullus barbatus* and *Merluccius merluccius*) collected in three different geographical sub-areas of the Mediterranean Sea. Mar Pollut Bull 140:129–137
2. Alberghini L, Truant A, Santonicola S, Colavita G, Giaccone V (2023) Microplastics in fish and fishery products and risks for human health: a review. Int J Environ Res Public Health 20:789. https://doi.org/10.3390/ijerph20010789
3. Rodríguez-Romeu O, Constenla M, Carrassón M, Campoy-Quiles M, Soler-Membrives A (2020) Are anthropogenic fibres a real problem for red mullets (*Mullus barbatus*) from the NW Mediterranean? Sci Total Environ 733:139336
4. Alomar C, Sureda A, Capó X, Guijarro B, Tejada S, Deudero S (2017) Microplastic ingestion by *Mullus surmuletus* Linnaeus, 1758 fish and its potential for causing oxidative stress. Environ Res 159:135–142
5. Capillo G, Savoca S, Panarello G, Mancuso M, Branca C, Romano V, Spanò N (2020) Quali-quantitative analysis of plastics and synthetic microfibers found in demersal species from Southern Tyrrhenian Sea (Central Mediterranean). Mar Pollut Bull 150:110596
6. Avio CG, Gorbi S, Regoli F (2015) Experimental development of a new protocol for extraction and characterization of microplastics in fish tissues: first observations in commercial species from Adriatic Sea. Mar Environ Res 111:18–26
7. Avio CG, Pittura L, d'Errico G, Abel S, Amorello S, Marino G, Regoli F (2020) Distribution and characterization of microplastic particles and textile microfibers in Adriatic food webs: general insights for biomonitoring strategies. Environ Pollut 258:113766
8. Güven O, Gökdağ K, Jovanović B, Kıdeyş AE (2017) Microplastic litter composition of the Turkish territorial waters of the Mediterranean Sea, and its occurrence in the gastrointestinal tract of fish. Environ Pollut 223:286–294
9. Neves D, Sobral P, Ferreira JL, Pereira T (2015) Ingestion of microplastics by commercial fish off the Portuguese coast. Mar Pollut Bull 101:119–126
10. Bellas J, Martínez-Armental J, Martínez-Cámara A, Besada V, Martínez-Gómez C (2016) Ingestion of microplastics by demersal fish from the Spanish Atlantic and Mediterranean coasts. Mar Pollut Bull 109:55–60
11. Atamanalp M, Köktürk M, Uçar A, Duyar HA, Özdemir S, Parlak V, Alak G (2021) Microplastics in tissues (brain, gill, muscle and gastrointestinal) of *Mullus barbatus* and *Alosa immaculata*. Bull Environ Contam Toxicol 81:460–469
12. Santonicola S, Volgare M, Cocca M, Dorigato G, Giaccone V, Colavita G (2023) Impact of fibrous microplastic pollution on commercial seafood and consumer health: a review. Animals 13:1736. https://doi.org/10.3390/ani13111736
13. Karami A, Golieskardi A, Ho YB, Larat V, Salamatinia B (2017) Microplastics in eviscerated flesh and excised organs of dried fish. Sci Rep 7:1–9
14. Daniel DB, Ashraf PM, Thomas SN (2020) Microplastics in the edible and inedible tissues of pelagic fishes sold for human consumption in Kerala, India. Environ Pollut 266:115365
15. Digka N, Tsangaris C, Torre M, Anastasopoulou A, Zeri C (2018) Microplastics in mussels and fish from the Northern Ionian Sea. Mar Pollut Bull 135:30–40
16. Santonicola S, Volgare M, Di Pace E, Cocca M, Mercogliano R, Colavita G (2021) Occurrence of potential plastic microfibers in mussels and anchovies sold for human consumption: preliminary results. Ital J Food Saf 10
17. Santonicola S, Volgare M, Di Pace E, Mercogliano R, Cocca M, Raimo G, Colavita G (2023) Research and characterization of fibrous microplastics and natural microfibers in pelagic and benthic fish species of commercial interest. Ital J Food Saf 12(1)
18. Barboza LGA, Cunha SC, Monteiro C, Fernandes JO, Guilhermino L (2020) Bisphenol A and its analogs in muscle and liver of fish from the North East Atlantic Ocean in relation to microplastic contamination. Exposure and risk to human consumers. J Hazard Mater 393:122419

19. Carbonara P, Intini S, Modugno E, Maradonna F, Spedicato MT, Lembo G, Carnevali O (2015) Reproductive biology characteristics of red mullet (*Mullus barbatus* L., 1758) in Southern Adriatic Sea and management implications. Aquat Living Resour 28:21–31

Chapter 27
Temporal Variability of Microplastics and Their Attached Community Along the Campania Coast

Vincenzo Donnarumma, Fabio D'Agostino, Roberta Piredda, and Raffaella Casotti

27.1 Introduction

Coastal marine ecosystems are continuously exposed to anthropogenic impacts such as densely urbanized areas, tourism, recreational and commercial fishing activities among others. This constant exposure of coastal areas to human activities increases the risk of mismanaged waste entering the marine environment, especially plastic [1]. As microplastics (MPs) are now widely recognized as marine pollutants, it is fundamental to define common methodologies to quantify their presence in all the environmental matrices, seawater among these. Despite the well-known subtropical ocean gyres acting as accumulation areas for floating plastic debris, the spatial distribution of these particles in the Mediterranean Sea is also affected by the variability of the surface circulation which hampers the formation of stable retention zones [2]. In addition, temporal variability has also been often recorded, even within same seasons. However, the potential influence of the nearshore dynamics in the spatio-temporal understanding of floating plastic and its potential effects on the retention of

V. Donnarumma (✉) · R. Casotti
EMI Department, Stazione Zoologica "Anton Dohrn", Villa Comunale, Napoli, Italy
e-mail: vincenzo.donnarumma@sp.ismar.cnr.it

V. Donnarumma
Institute for Marine Sciences (ISMAR), Consiglio Nazionale Delle Ricerche, Pozzuolo Di Lerici (SP), Italy

F. D'Agostino
Institute for Coastal and Marine Environment (IAMC), Detached Units of Capo Granitola, Consiglio Nazionale Delle Ricerche, Mazara del Vallo (TP), Italy

R. Piredda
Veterinary Medicine Department, Università Degli Studi Di Bari Aldo Moro, Bari, Italy

M. Cocca et al. (eds.), *Proceedings of the 3rd International Conference on Microplastic Pollution in the Mediterranean Sea*, Springer Water,
https://doi.org/10.1007/978-3-031-34455-8_28

plastics is a current knowledge gap. Hence sampling campaign data and marine circulation data are fundamental in the microplastic pollution study [3]. Similar trends of variability are linked to the community of microbes colonizing the surface of MPs, the so called "Plastisphere" [4]. When a piece of plastic enters the ocean, microbes colonize it within hours, and from there on different factors contribute on shaping the community.

Geographic location and environmental parameters appear to be the primary influences shaping Plastisphere communities, but studies using environmentally collected samples are rare [5]. Studies show that ocean-basin scale [6], coastal areas [7] and riverine environments [8] differences exist. These researches demonstrate that there is a clear need to consider variability along spatial and temporal gradients, considering the movement of plastics from lakes to rivers onward to the sea, as well as the original nature of the polymer, to better understand dynamic variation in the structure, diversity and ecological roles of plastisphere microbial communities with marine plastic pollutants.

Very few info on MPs in the Gulf of Naples is available, for example a technical report from CNR-ISMAR from a sampling cruise in summer 2017, indicates 0.26 MPs m^{-3} at Punta Campanella and 3.56 close to the city of Portici, our st.1 [9]. Floating MP concentrations, distribution, and characteristics of the associated microbial plastisphere were investigated with the aim of assessing their space and time distribution, identifying possible accumulation areas, assessing putative differences between polymers in terms of attached bacteria, trying to identify common members of the microbial plastisphere, to be interpreted as obligate plastic-associated organisms. This research is part of the PhD thesis of Vincenzo Donnarumma (Open University-Stazione Zoologica "Anton Dohrn" [10]).

27.2 Sampling Technique and Storage

MPs were collected in January and July 2018 (MP18A and MP18B, respectively), August and September 2019 (MP19A and MP19B, respectively) and January 2020 (MP20A) at three stations in the Gulf of Naples. The three sampling sites were, respectively, in front of the small town of Portici (st. 1, 0.5 nm from the coast), in the middle of the bay at the Long Term Ecological Research station "MareChiara" (st. 2, 2 NM away from the coast) and at the Marine Protected Area of Gaiola (st.3, 0.5 nm from the coast). MPs were collected using a 0.6 × 0.16 m manta net with a 333-μm mesh size (Oceomic, Spain). In the center of the opening of the manta net, a flowmeter was placed to estimate the volume of water sampled. The net was towed usually for 30 min at an average speed of 2 knots or less. Once removed, the cod-end content was sieved through two sieves of 5000 and 300 μm in sequence in order to isolate the material in the desired size range. In order to characterize the plastisphere community, the largest plastic pieces within 300 and 5000 μm of plastic were removed with sterile forceps, rinsed with 0.22 μm filter-sterilized seawater and cut into three sub-pieces for DNA extraction and chemical composition analysis by Fourier-Transform Infrared Spectroscopy (FT-IR), respectively. The remaining pieces in the smaller sieve, together with all organic and inorganic material, including zooplankton and other organisms retained, were then poured into a glass container, fixed with 70% ethanol and stored at 4 °C for subsequent analysis. All the equipment was cleaned, and all items were stored in clean covered (during work) or sealed (for storage) Petri dishes. Plastic for downstream DNA analysis were immediately placed in 1.5 ml Eppendorf tubes filled with Puregene lysis buffer (Qiagen, Valencia, CA) and frozen at −20 °C. Finally, plastic samples for FT-IR were placed in individual 1.5 ml Eppendorf tubes. The sampled area and the water volume filtered during each trawl were calculated using the frame dimensions and the tow distance (derived from the flowmeter readings). MP concentrations were expressed as particles m^{-3}.

27.2.1 Microplastics Counting and Classification

Once in the lab, the samples were poured into a Petri dish and analysed with the use of a stereomicroscope (Leica CLS 150 XE, using a 20–80 × zoom) for the count and classification of microplastic particles. MP particles were distributed into six categories according to their visual features, four of them belonging to secondary microplastics and two to primary ones. The most abundant were usually secondary microplastics divided into fragments, films, foams and filaments. As for primary microplastics there were two types: pellets and granules.

27.2.2 DNA Extraction, Sequencing and Bioinformatic Analyses

DNA was extracted from either single plastic pieces using a modified bead-beating approach in combination with the Puregene Tissue DNA extraction kit (Qiagen, Valencia, CA). DNA was quantified with Nanodrop spectrophotometer. The extracted DNA was sent to the CGBM Sequencing Facility of the Dalhousie University, Canada, for 2 × 300 paired-end (PE) Illumina MiSeq sequencing using the primers 515F-Y (5′-GTGYCAGCMGCCGCGGTAA) and 926R (5′-CCGYCAATTYMTTTRAGTTT) of the 16S rDNA gene [9]. The raw data in fastq format were preprocessed by the Bioinforma service at SZN using the pipeline implemented in QIIME 2. To enable comparison between samples, the cleaned dataset was randomly subsampled to the smallest number of sequences in one sample, using the function 'rarefy' in vegan R package. Bray–Curtis Dissimilarity matrix was computed (vegdist in vegan) and used for subsequent ordination analyses. Non-metric multidimensional scaling (NMDS) was performed using the metaNMDS function in vegan. Venn diagrams were obtained using the Venny open-source online website.

27.3 Results and Discussions: *Microplastics Characteristics*

MP concentrations ranged around one order of magnitude, from 1.12 to 12.08 MP m^{-3}. Highest concentrations found were during summer 2019 (both August and September). The lowest during the two winter period times (January 2018 and 2020, Table 27.1). Throughout the years monitored, MPs ranging between 1000 and 5000 μm, called large MPs, accounted for 79.7% of total MPs. Small MPs (ranging among 300–1000 μm) accounted for 16.2% of total MPs. The other MPs were all less than 333 μm, our minimum size limit of the mesh size of the manta-net used. This confirm a high efficiency of the device used, since less than 4% of the MPs were smaller than the manta net size. Three out of five times large MPs accounted for more than 80% contribution of MP most common size fraction, and the only times that this did not happen is when the highest MP concentrations were found (August and September 2019).

Fragments were the most common type of MPs found, averaging 63.4% of the total MP found, followed by films (25.2%), filaments (6.9%) and foams (4.4%). The remaining 0.2% accounted for the so-called primary MPs, clearly not affecting the Gulf of Naples, in granules or pellets. Some exceptions were found when in January 2020, very similar values were found comparing fragments and films. In the same season—but different year, 2018—was found a high contribution of films. As for MP colour during two samplings in 2018, coloured MPs dominated, both in January and July. A true shift of this trend was found from the year 2019. Indeed before, coloured MPs were more abundant, but from that year on, transparent MPs dominated. On the contrary, white has always been the least found colour, even though, noteworthy is the peak found in st.3 in September 2019. Data showed here are in agreement with a rising concern regarding the variability that comes with the assessment of MP pollution when it finds different environmental conditions in particular areas, highlighting the limits of single-spot surveys and the need of continuous samplings, especially in coastal areas.

Table 27.1 MP concentration related to the different stations of the Gulf of Naples (st.1, 2 and 3) at the different samplings. Data are in MP m^{-3}

	St.1	St.2	St.3
Jan-18	1.12	1.78	1.27
Jul-18	4.66	1.60	–
Aug-19	9.46	8.39	1.29
Sep-19	12.08	9.89	5.32
Jan-20	1.13	1.42	1.78
Average	5.69	4.61	2.41
Dev.st	4.94	4.16	1.95

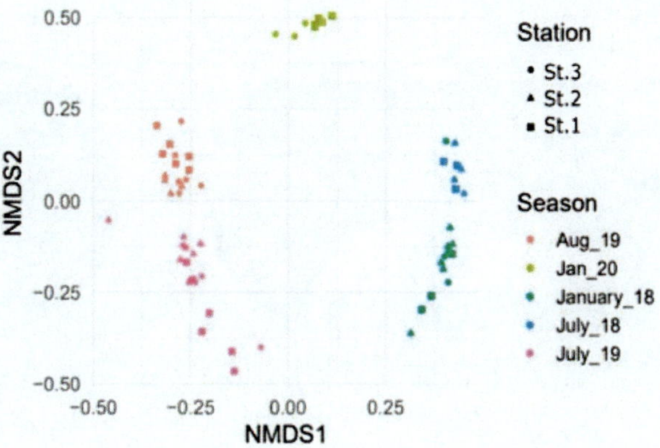

Fig. 27.1 Temporal non-metric multidimensional scaling (NMDS) representation of Bray–Curtis dissimilarity matrix of the attached prokaryotic communities (stress 9.5 e−05)

27.3.1 Results and Discussions: Plastisphere

75 MPs pieces were retrieved for parallel plastisphere characterization in the Gulf of Naples. 49 Polypropylene (PE), 13 Polypropylene (PP), 5 Polystyrene (PS), 2 Ethylene vinyl acetate (EVA), 2 Poly(Ethylene:Propylene:Diene) (PEPD), 2 Waxes (WAX), 1 Polyurethane (PU), 1 Cellophane. 97.3% of the polymers recognized in this work were positively buoyant (PE, PP, expanded PS, EVA, PEPD, Wax) while only PU and Cellophane were negatively buoyant. PE was the only polymer found in all samples. In order to better characterize the prokaryotes composing the microbial plastisphere, 74 MPs were used to extract DNA and perform 16S rRNA sequencing. Non-metric multidimensional scaling (NMDS) based on Bray–Curtis dissimilarities was performed in order to highlight patterns in gradients related to community assessments. The NMDS revealed that the global pattern of prokaryotic diversity was not explained by a spatial trend but rather by a temporal one, since dates clustered together all stations from the same sampling date (Fig. 27.1).

In order to investigate the confirm the high variability in terms of spatial parameters, were analysed OTUs displayed in Venn diagrams. No OTUs were shared between the years, and the highest percentages of the Venn diagrams were all related to exclusive OTUs. The only shared OTUs higher than 0 (0.9%) were among the years 2019 and 2020 (Fig. 27.2a). As PE was the only polymer shared among all the seasons retrieved, the PE core microbiome was analyzed for shared and unique OTUs throughout the years. Similar results for PE are showed in Fig. 27.2b, with only 6 OTUs shared between 2019 and 2020. Molecular results presented showed that there is a clear signal related to the temporal variability linked to the attached community structure in the Gulf of Napoli. This is likely related to the origin of the samples collected, as they were sampled from the environment, and not, as the majority of studies of plastisphere, from incubation experiments.

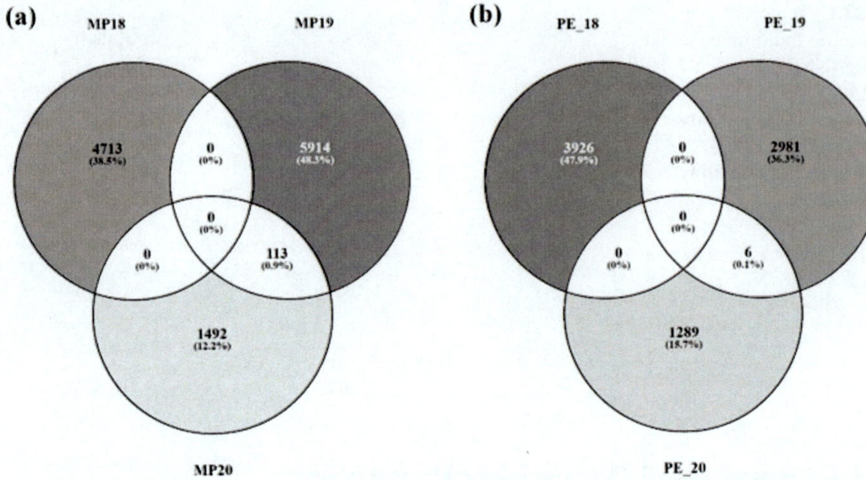

Fig. 27.2 Venn diagrams showing shared and exclusive OTUs among different polymer associated communities (**a**) and PE (**b**) at the different sampling dates. Numbers shown are actual OTUs and percentage of dataset contribution

27.4 Conclusions

In conclusion, the data presented highlight the importance of repeated sampling to assess MP pollution and its attached community in coastal areas because of complex circulation and multiple terrestrial discharges. Variability in the Gulf of Naples is very high, both in terms of total concentrations and community composition of attached microbes. The plastisphere community data presented here underlined again the importance of temporal variability and environmental parameters shaping the microbial community, as reported by other studies [6, 11]. Further studies should be aimed at functional measurements of microbial activity of MPs, so to better understand their role as new habitats and also their potential threat to the environment and the final consumers [12].

References

1. Compa M, Alomar C, Mourre B, March D, Tintoré J, Deudero S (2020) Nearshore spatio-temporal sea surface trawls of plastic debris in the Balearic Islands. Mar Environ Res 158:104945
2. Ruiz-Orejón LF, Mourre B, Sardá R, Tintoré J, Ramis-Pujol J (2019) Quarterly variability of floating plastic debris in the marine protected area of the Menorca Channel (Spain). Environ Pollut 252:1742–1754
3. Pini A, Tomassetti P, Matiddi M, de Lucia GA, Camedda A, Zampetti G, Lattanzi C, Leuzzi G, Monti P (2018) Microplastic samplings and inverse trajectory recognition in the Mediterranean

Sea. In: 2018 IEEE international workshop on metrology for the sea; learning to measure sea health parameters (MetroSea). IEEE, pp 115–119
4. Zettler ER, Mincer TJ, Amaral-Zettler LA (2013) Life in the "Plastisphere": microbial communities on plastic marine debris. Environ Sci Technol 47:7137–7146. https://doi.org/10.1021/es401288x
5. Amaral-Zettler LA, Zettler ER, Mincer TJ (2020) Ecology of the plastisphere. Nat Rev Microbiol 18:139–151. https://doi.org/10.1038/s41579-019-0308-0
6. Amaral-Zettler LA, Zettler ER, Slikas B, Boyd GD, Melvin DW, Morrall CE, Proskurowski G, Mincer TJ (2015) The biogeography of the Plastisphere: implications for policy. Front Ecol Environ 13:541–546. https://doi.org/10.1890/150017
7. Amaral-Zettler LA, Ballerini T, Zettler ER, Asbun AA, Adame A, Casotti R, Dumontet B, Donnarumma V, Engelmann JC, Frère L, Mansui J, Philippon M, Pietrelli L, Sighicelli M (2021) Diversity and predicted inter- and intra-domain interactions in the Mediterranean Plastisphere. Environ Pollut 286:117439. https://doi.org/10.1016/j.envpol.2021.117439
8. Oberbeckmann S, Loeder MGJ, Gerdts G, Osborn AM (2014) Spatial and seasonal variation in diversity and structure of microbial biofilms on marine plastics in Northern European waters. FEMS Microbiol Ecol 90:478–492. https://doi.org/10.1111/1574-6941.12409
9. Mezzelani M, Avio CG, Nardi A, Regoli F (2017) Microplastic investigation in water and trophic chain along the Italian coast 06/24/2017–07/15/2017
10. Donnarumma V (2022) Microplastics in the marine environment and the characterization of their attached microbial communities. Ph.D. thesis The Open University
11. Parada AE, Needham DM, Fuhrman JA (2016) Every base matters: assessing small subunit rRNA primers for marine microbiomes with mock communities, time series and global field samples. Environ Microbiol 18(5):1403–1414
12. Oberbeckmann S, Labrenz M (2020) Marine microbial assemblages on microplastics: diversity, adaptation, and role in degradation. Ann Rev Mar Sci 12:209–232. https://doi.org/10.1146/annurev-marine-010419-010633

Chapter 28
The Vertical Distribution of Riverine Microplastics: The Role of Turbulence

Hadeel Al-Zawaidah, Bart Vermeulen, and Kryss Waldschläger

28.1 Introduction

Owning to plastics high durability and resistance to degradation [1] alongside the high production rates [1–3], plastic pollution and its impacts have reached every aquatic and terrestrial environment [4–7]. In the water and sediments of aquatic systems, sampling campaigns confirmed and documented a plethora of microplastic (<5 mm) (e.g., [8–12]). Estimates suggest that the total amount of plastics immitted to the world oceans is approximately 4.8–12.7 MT [7], of which a high percentage is thought to reach the ocean through rivers [13, 14]. Hence, rivers are seen as a primary pathway for microplastics to marine environments. Recent observations further suggest that rivers represent a sink for plastic litter where it may be retained for prolonged periods before being released into the ocean [15]. Rivers can also act as a hosting environment for larger plastic litter (>5 mm) fragmenting into secondary microplastics. Once in rivers, microplastics can lead to a variety of ecological, economic, and hazardous impacts [16, 17]. Understanding the fate and transport of microplastics in riverine systems is, therefore, a crucial step toward establishing effective protocols to mitigate against these emerging pollutants. Accurate estimates of microplastics abundance and transport in rivers remain far from reach [18], however. Despite the abundance of water surface and sediments sampling campaigns (e.g., [19–21]), the record of microplastics occurrence along the water column is still in its infancy [22]. In fact, microplastics vertical distribution within the water column and the associated governing processes remain poorly understood [22, 23].

The general practice in estimating microplastics in rivers is through adopting a predefined distribution along the water column, hereafter a concentration depth

H. Al-Zawaidah (✉) · B. Vermeulen · K. Waldschläger
Hydrology & Quantitative Water Management Group, Wageningen University and Research, Wageningen, The Netherlands
e-mail: hadeel.alzawaidah@wur.nl

M. Cocca et al. (eds.), *Proceedings of the 3rd International Conference on Microplastic Pollution in the Mediterranean Sea*, Springer Water,
https://doi.org/10.1007/978-3-031-34455-8_29

profile, which is mainly linked to microplastics density. With the wide range of density covered by the plastic polymers (i.e., being positively, negatively, and neutrally buoyant) [24], plastic density assumptions vary among studies. As a result, microplastics were assumed to be either transported below the surface as bedload and suspended load [23, 25], surface load [13, 26], or neutral wash load regardless of the particle and flow field characteristics [27, 28]. On the contrary, the limited documentation of microplastics depth profiles supports microplastics transport along various levels of the water column [29–31]. Even further, the limited water column sampling campaigns landed on different conclusions, including but not limited to (i) the dominance of the surface load mode of transport with a limited suspension due to turbulence [32], (ii) a correlation between microplastic depth profile and flow conditions [31] and (iii) a correlation between microplastics density and depth profile [29].

These observations accompanied by the lack of understanding of other microplastics characteristics (i.e., size and shape) on their concentration depth profile, stress the need for further investigation to develop new models, where the governing mechanisms are taken into consideration [23]. In response, a process-based model based on the physical processes influencing microplastics concentration depth profile has been proposed [33]. The proposed model is developed from the Rouse model, hence employing principles previously adopted in sedimentological studies [34]. The concentration depth profile of microplastics is assumed to be governed by the interaction between gravity forces and turbulent mixing. Despite the extensive validation for sediments [34–37], model validation for microplastics remains missing. The primary parameters for the turbulent mixing process are the eddy viscosity and the material diffusivity [38, 39]. Their ratio (i.e., Prandtl-Schmidt number [40]), is a primary input for the Rouse model, yet it remains unknown for microplastics. Further, the Rouse model's capacity of representing real-life scenarios is challenged by its one-dimensional nature, reflecting simplified scenarios of turbulence. Examination of the impact of turbulence on the vertical distribution is, therefore, essential for the validation of the proposed Rouse model and the assessment of its potential and limitations. With this research project we aim to (i) better understand the concentration depth profile of microplastics within turbulence, (ii) validate and calibrate the Rouse model for microplastics, and (iii) explore the potential of computational numerical modelling for complex scenarios.

28.2 Methods

The project involves employing a combination of physical and numerical modelling. The physical modelling will allow for generating the necessary datasets, which will support a better understanding of the impact of turbulence on the vertical distributions and will help testing the Rouse model performance. The numerical modelling will allow for exploring extended scenarios of complex turbulence while potentially providing a prediction tool for the wide range of combinations of flow conditions and particle characteristics.

28.2.1 Physical Modelling

28.2.1.1 Experimental Setup

The FC 300 flume (5 m long with a cross-section of 450 mm × 300 mm) at the laboratory of water and sediments dynamics at Wageningen University is employed for the experimental setup. The flume allows for continuous and separate circulation of solid and liquid phases with an adjustable channel slope. The effective section is set within the middle 3 m of the flume length to minimise the inlet/outlet impacts. The experiment includes three series, mimicking three different flow conditions. Using the sediment circulation system, homogeneous microplastic particles are circulated (particles of the same size, shape, and density). For each run, microplastics supply rate is gradually increased, up to a point where the depth distribution (i.e., microplastics concentration) can be obtained. However, the upper limit of the supply rate is constrained by the limit for interparticle interactions [41], as such processes are unlikely in the light of microplastics concentration reported in rivers (see [42]). Three repetitions are carried out of each experimental run to assess the results' reproducibility. Once hydrodynamic equilibrium conditions are established (i.e., uniform flow) measurements are recorded using a particle image velocimetry setup (PIV) (Fig. 28.1).

In comparison to traditional approaches (e.g., acoustics devices and suction sampling), PIV systems represent a non-intrusive technique with no interference or alteration of the flow within the flume [43, 44]. The setup targets tracking two phases.

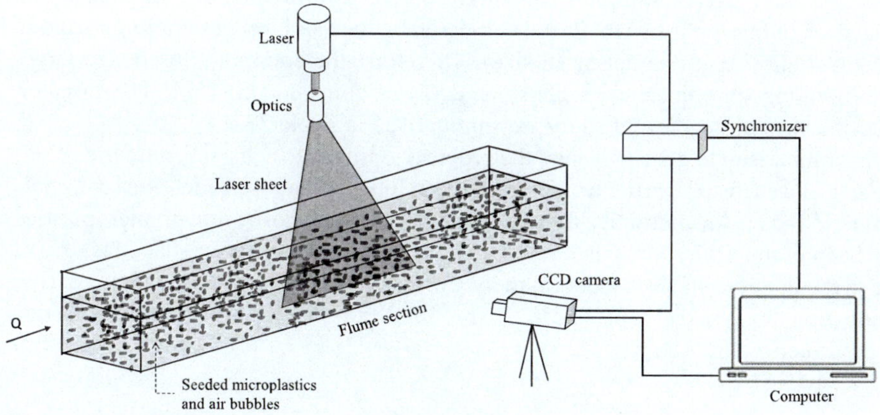

Fig. 28.1 Experimental setup of the physical modelling

Information about microplastic concentration depth profile and rate of transport are gathered by tracking microplastics (i.e., the solid phase), employing a Lagrangian analysis of the collected images. The tracer phase refers to the water flow monitored through a tracer (air bubbles in the case of this setup), which is analysed employing a Eulerian analysis to infer the flow characteristics (i.e., the instantaneous and mean of the three components of the flow velocity). The turbulent shear stresses and strain rates needed to determine the eddy viscosity are obtained from the Eulerian analysis, while the concentration depth profiles and microplastics diffusivity are determined by the Lagrangian analysis. Finally, the ratio between the eddy viscosity and particle diffusivity (i.e., Prandtl-Schmidt number) is obtained.

The image analysis is carried out employing the open source PIVview software [45]. As the images are collected instantaneously, a pre-processing of the collected images is needed to disintegrate the two phases prior to analysis. Different methods are adopted in literature to separate the tracer from the solid phase (e.g., spot size [43], grey level [46], and colour/fluorescence [47]. The spot size approach is deemed suitable for the setup as it reduces the complexity of the needed PIV setup.

28.2.1.2 Material

For plastic litter (including microplastics and macroplastics), the shape, size and density of the objects are key factors determining their transport in rivers [48]. A systematic variation of microplastics characteristics (i.e., shape, size, and density) is conducted to assess the impact of each property individually. Nevertheless, the selection of a representative set of microplastics, mimicking riverine microplastic, remains challenging due to the diversity of the material properties. Previous research on microplastics in riverine systems is, therefore, used to guide the selection of the microplastics for the experiment. The work of Liu et al. [22] and Kumar et al. [49] offers an insight into the dominant shapes and polymer types (i.e., density) of microplastics based on sampling campaigns in rivers worldwide. Inferring the dominant sizes of microplastics in rivers is hindered by the variation in sampling methods among studies (e.g., the net mesh size). Therefore, the commonly used size range incorporating microplastics is considered (i.e., 5 mm–1 μm) [50]. The targeted microplastics in the present study are summarised in Table 28.1.

In comparison to density, where the property is linked to a discrete numeric value, the size and shape properties are connected and linked to multiple identification challenges [51–53]. An alternative to the commonly used categorization of microplastics has been proposed by Melkebeke et al. [54], employing descriptive shape factors. In the present study, shape factors (namely Corey shape factor, sphericity, circularity, elongation, flatness and aspect ratio) are reported and examined for the microplastics in the experimental setup.

Table 28.1 Microplastics for the experimental set-up (selected based on Liu et al. [22], Duis and Coors [55] and Kumar et al. [49])

Polymer	Density (g cm^{-3})	Shapes
Polyethylene (PE)	0.89–0.98	Fibres, fragments, films, foams, pellets, and micro-beads
Expanded polystyrene (Styrofoam) (EPS)	0.01–0.04	
Polystyrene (PS)	1.04–1.1	
Polypropyleen (PP)	0.83–0.92	
PEST (Polyester)/ Poly(ethersulfone) (PES)	1.24–2.3	
Polyethylene terephthalate (PET)	0.96–1.45	
Polyvinyl acetate (PVA)	1.2–1.3	
Polyvinylchloride (PVC)	1.16–1.58	

28.2.2 Numerical Modelling

Due to feasibility and technical limitations of the physical modelling, the range of particles and flow characteristics is limited. Reynolds-averaged Euler–Lagrange formulations, previously explored for sediment transport, are planned for the numerical modelling exercise. The open-source coupled CFD-DEM scheme CFDEM-EIM software is adopted, where previous applications in sedimentological studies can offer guidance [56, 57]. In the present research, the intergranular interaction is not included considering microplastic concentrations in rivers. Yet, the discrete element method built within the software offers room for model adaptation to higher concentrations, where inter-particle interactions might be relevant. Datasets obtained from the physical model feed into the model validation, where the model capacity to regenerate lab observations can be tested. Once the model is validated, it can be used to explore combinations of flow conditions and particle characteristics beyond the laboratory experiment.

An additional step is to assess the performance of the Rouse model proposed by Cowger et al. [33] for the concentration depth profile of microplastics against observed concentration depth profiles at the lab and the model output.

28.3 Anticipated Outcomes

The anticipated project results will be useful to amend the limitations of water column sampling campaigns and could help understand the driving factors causing the variations between the findings of the documented water column observations. The unique gathered dataset can be used to parameterise the influence of turbulence on microplastics transport and vertical distribution. Further, the numerical model can produce a prediction tool for microplastics occurrence and concentration in rivers. Ultimately, the result translates into a better quantification of microplastics occurrence in rivers, which could lead to (i) more efficient and accurate sampling practices, (ii) better estimation of the exposure dose along the water column for toxicology studies, and (iii) targeted and efficient mitigation measures for microplastics in rivers.

References

1. Andrady AL (2003) Plastics and the environment. Ohn Wiley and Sons, Hoboken, NJ
2. Harrison RM, Hester RE (ed) (2019) Plastics and the environment. Royal Society of Chemistry
3. Baur E, Brinkmann S, Osswald T, Rudolph N, Schmachtenberg E (2013) Saechtling Kunststoff Taschenbuch, 31st Editi. Hanser Publishers, Munich
4. Geyer R, Jambeck JR, Law KL (2017) Production, use, and fate of all plastics ever made. Sci Adv 3:25
5. Lebreton L, Egger M, Slat B (2019) A global mass budget for positively buoyant macroplastic debris in the ocean. Sci Rep 9:1
6. Thompson RC, Olsen Y, Mitchell RP, Davis A, Rowland SJ, John AWG, McGonigle D, Russell AE (2004) Lost at sea: where is all the plastic? Science 304(80):838
7. Jambeck JR, Geyer R, Wilcox C, Siegler TR, Perryman M, Andrady A, Narayan R, Law KL (2015) Plastic waste inputs from land into the ocean. Science 347(80):768
8. Browne MA, Crump P, Niven SJ, Teuten E, Tonkin A, Galloway T, Thompson R (2011) Accumulation of microplastic on shorelines woldwide: sources and sinks. Environ Sci Technol 45:9175
9. Driedger AGJ, Dürr HH, Mitchell K, Van Cappellen P (2015) Plastic debris in the Laurentian Great Lakes: a review. J Great Lakes Res 41:9
10. Koelmans AA, Mohamed Nor NH, Hermsen E, Kooi M, Mintenig SM, De France J (2019) Microplastics in freshwaters and drinking water: critical review and assessment of data quality. Water Res 155(410)
11. Rhodes CJ (2018) Plastic pollution and potential solutions. Sci Prog 101:207
12. Sundt P, Schulze PE, Syversen F (2014) Sources of microplastic-pollution to the marine environment. Mepex for the Norwegian Environment Agency
13. Lebreton LCM, Van Der Zwet J, Damsteeg JW, Slat B, Andrady A, Reisser J (2017) River plastic emissions to the world's oceans. Nat Commun 8:1
14. Andrady AL (2011) Microplastics in the marine environment. Mar Pollut Bull 62:1596
15. van Emmerik T, Mellink Y, Hauk R, Waldschläger K, Schreyers L (2022) Rivers as plastic reservoirs. Front Water 3
16. Du H, Xie Y, Wang J (2022) Environmental impacts of microplastics on fishery products: an overview. Gondwana Res 108:213
17. Koelmans AA, Redondo-Hasselerharm PE, Nor NHM, de Ruijter VN, Mintenig SM, Kooi M (2022) Risk assessment of microplastic particles. Nat Rev Mater 0123456789
18. Barcelo D, Pico Y (2020) Case studies of macro- and microplastics pollution in coastal waters and rivers: is there a solution with new removal technologies and policy actions? Case Stud Chem Environ Eng 2:100019
19. Adomat Y, Grischek T (2021) Sampling and processing methods of microplastics in river sediments—a review. Sci Total Environ 758:143691
20. Kataoka T, Nihei Y, Kudou K, Hinata H (2019) Assessment of the sources and inflow processes of microplastics in the river environments of Japan. Environ Pollut 244:958

21. Luo W, Su L, Craig NJ, Du F, Wu C, Shi H (2019) Comparison of microplastic pollution in different water bodies from urban creeks to coastal waters. Environ Pollut 246:174
22. Liu K, Courtene-jones W, Wang X, Song Z, Wei N (2020) Elucidating the vertical transport of microplastics in the water column: a review of sampling methodologies and distributions. Water Res 186:116403
23. Besseling E, Quik JTK, Sun M, Koelmans AA (2017) Fate of nano- and microplastic in freshwater systems: a modeling study. Environ Pollut 220:540
24. Ballent A, Pando S, Purser A, Juliano MF, Thomsen L (2013) Modelled transport of benthic marine microplastic pollution in the Nazaré canyon. Biogeosciences 10:7957
25. Nizzetto L, Bussi G, Futter MN, Butterfield D, Whitehead PG (2016) A theoretical assessment of microplastic transport in river catchments and their retention by soils and river sediments. Environ Sci Process Impacts 18:1050
26. Miller RZ, Watts AJR, Winslow BO, Galloway TS, Barrows APW (2017) Mountains to the sea: river study of plastic and non-plastic microfiber pollution in the Northeast USA. Mar Pollut Bull 124:245
27. Schmidt C, Krauth T, Wagner S (2017) Export of plastic debris by rivers into the sea. Environ Sci Technol 51:12246
28. Lechner A, Keckeis H, Lumesberger-Loisl F, Zens B, Krusch R, Tritthart M, Glas M, Schludermann E (2014) The Danube so colourful: a potpourri of plastic litter outnumbers fish larvae in Europe's second largest river. Environ Pollut 188:177
29. Lenaker PL, Baldwin AK, Corsi SR, Mason SA, Reneau PC, Scott JW (2019) Vertical distribution of microplastics in the water column and surficial sediment from the Milwaukee River Basin to lake Michigan. Environ Sci Technol 53:12227
30. Morritt D, Stefanoudis PV, Pearce D, Crimmen OA, Clark PF (2014) Plastic in the thames: a river runs through it. Mar Pollut Bull 78:196
31. Haberstroh CJ, Arias ME, Yin Z, Wang MC (2020) Effects of hydrodynamics on the cross-sectional distribution and transport of plastic in an Urban Coastal River. Water Environ Res 1386
32. Lestari P, Trihadiningrum Y, Wijaya BA, Yunus KA, Firdaus M (2020) Distribution of microplastics in Surabaya River, Indonesia. Sci Total Environ 726:138560
33. Cowger W, Gray AB, Guilinger JJ, Fong B, Waldschläger K (2021) Concentration depth profiles of microplastic particles in river flow and implications for surface sampling. Environ Sci Technol 55:6032
34. de Leeuw J, Lamb MP, Parker G, Moodie AJ, Haught D, Venditti JG, Nittrouer JA (2020) Entrainment and suspension of sand and gravel. Earth Surf Dyn 8:485
35. Lamb MP, de Leeuw J, Fischer WW, Moodie AJ, Venditti JG, Nittrouer JA, Haught D, Parker G (2020) Mud in rivers transported as flocculated and suspended bed material. Nat Geosci 13:566
36. Armijos E et al (2017) Measuring and modeling vertical gradients in suspended sediments in the Solimões/Amazon River. Hydrol Process 31:654
37. Eggenhuisen JT, Tilston MC, Leeuw J, Pohl F, Cartigny MJB (2020) Turbulent diffusion modelling of sediment in turbidity currents: an experimental validation of the rouse approach. Depos Rec 6:203
38. Graf WH, Cellino M (2002) Suspension flows in open channels; experimental study. J Hydraul Res 40:435
39. Karimpour F, Venayagamoorthy SK (2014) A simple turbulence model for stably stratified wall-bounded flows. J Geophys Res Ocean 119:870
40. Sassi MG, Hoitink AJF, Vermeulen B (2013) Quantified turbulent diffusion of suspended sediment using acoustic doppler current profilers. Geophys Res Lett 40:5692
41. Vowinckel B (2021) Incorporating grain-scale processes in macroscopic sediment transport models. Acta Mech 232:2023
42. Hurley R, Woodward J, Rothwell JJ (2018) Microplastic contamination of river beds significantly reduced by catchment-wide flooding. Nat Geosci 11:251

43. Nezu I, Sanjou M (2011) PIV and PTV measurements in hydro-sciences with focus on turbulent open-channel flows. J. Hydro-Environ Res 5:215
44. Chen W, Chang K-C (2020) Dev PIV Meas Tech Turbul Flow Laden Binary-Size Part Groups 37:161
45. PIVTEC GmbH (2012) PIV view User Manual, Version 3.5 (PIVTEC GmbH)
46. Shi B, Wei J, Pang M (2015) Modif cross-Correl algorithm PIV image process part fluid two phase flow 45:105
47. Elhimer M, Praud O, Marchal M, Cazin S, Bazile RS (2016) Simultaneous PIV/PTV velocimetry technique in a turbulent particle-laden flow; Elhimer M, Praud O, Marchal M, Cazin S, Bazile R (2017) J Vis 19(1)
48. Schwarz AE, Ligthart TN, Boukris E, van Harmelen T (2019) Sources, transport, and accumulation of different types of plastic litter in aquatic environments: a review study. Mar Pollut Bull 143:92
49. Kumar R, Sharma P, Manna C, Jain M (2021) Abundance, interaction, ingestion, ecological concerns, and mitigation policies of microplastic pollution in riverine ecosystem: a review. Sci Total Environ 782:146695
50. Waldschläger K et al (2022) Learning from natural sediments to tackle microplastics challenges: a multidisciplinary perspective. Earth-Sci Rev 228:104021
51. Kooi M, Koelmans AA (2019) Simplifying microplastic via continuous probability distributions for size, shape, and density. Environ Sci Technol Lett 6:551
52. Hartmann NB et al (2019) Are we speaking the same language? recommendations for a definition and categorization framework for plastic debris. Environ Sci Technol 53:1039
53. Waldschläger K et al. (2022) Learning from natural sediments to tackle microplastics challenges: a multidisciplinary perspective. Earth-Sci Rev 104021
54. Van Melkebeke M, Janssen C, De Meester S (2020) Characteristics and sinking behavior of typical microplastics including the potential effect of biofouling: implications for remediation
55. Duis K, Coors A (2016) Microplastics in the aquatic and terrestrial environment: sources (with a specific focus on personal care products), fate and effects. Environ Sci Eur 28:1
56. Goniva C, Kloss C, Deen NG, Kuipers JAM, Pirker S (2012) Particuology influence of rolling friction on single spout fluidized bed simulation. Particuology 10:582
57. Cheng Z (2018) Eddy interaction model for turbulent suspension in Reynolds-averaged Euler-Lagrange simulations of steady sheet flow. Adv Water Resour 111:435

Chapter 29
Degradation of Biodegradable Plastic in Marine Environment

Thomas Viel, Roberto Avolio, Maria Emanuela Errico, Davide Caramiello, Giovanni Libralato, Alberto Macina, Loredana Manfra, Valerio Zupo, Maria Costantini, and Mariacristina Cocca

Abstract In the recent years, a growing attention has addressed to compostable and biodegradable polymers, in order to reduce plastic pollution. arding these latter, very few data on their biodegradability and toxicity in the marine environment are available as well as on their impact on marine invertebrates. In this study, we investigated the behaviour and degradation of five biodegradable polymers (BPs): Poly(butylene succinate) (PBS); Poly(butylene succinate-co-adipate) (PBSA); Polycaprolactone (PCL); Polyhydroxy butyrate (PHB); Polylactic acid (PLA) in marine environment.

T. Viel · G. Libralato
Department of Biology, University of Naples Federico II, Naples, Italy

T. Viel · R. Avolio · M. E. Errico · M. Cocca (✉)
Institute for Polymers Composites and Biomaterials, National Research Council of Italy, Pozzuoli, Italy
e-mail: maraicrsitina.cocca@cnr.it

T. Viel · D. Caramiello · G. Libralato · L. Manfra · M. Costantini
Department of Ecosustainable Marine Biotechnology, Stazione Zoologica Anton Dohrn, Villa Comunale, 80121 Napoli, Italy

D. Caramiello · A. Macina
Department of Research Infrastructures for Marine Biological Resources, Marine Organisms, Stazione Zoologica Anton Dohrn, Naples, Italy

L. Manfra
Institute for Environmental Protection and Research (ISPRA), Via Vitaliano Brancati 48, 00144 Rome, Italy

V. Zupo
Department of Ecosustainable Marine Biotechnology, Ischia Marine Centre, Stazione Zoologica Anton Dohrn, 80077 Ischia, Italy

M. Cocca et al. (eds.), *Proceedings of the 3rd International Conference on Microplastic Pollution in the Mediterranean Sea*, Springer Water,
https://doi.org/10.1007/978-3-031-34455-8_30

29.1 Introduction

Plastic are materials that are an integral part of our lives. It is estimated that the massive production of this material was around 6,300 million tonnes in 2015 and that around 8–14 million tonnes enter the ocean each year. In response to this problem, EU strategies have been put in place to reduce the release of plastics into the oceans, such as promoting recycling, reuse of plastics and the use of biodegradable plastics [1]. The most common plastics found in the aquatic environments are commonly used polymers such as high-density polyethylene (HDPE), low-density polyethylene (LDPE), polyvinyl chloride (PVC), polystyrene (PS), polypropylene (PP) and polyethylene terephthalate (PET) [2]. In the recent years, a growing attention has addressed to compostable and biodegradable polymers, in order to reduce plastic pollution. Regarding these latter, very few data on their biodegradability and toxicity in the marine environment are available as well as on their impact on marine invertebrates [1, 3]. In this scenario, there is an increasing attention towards biodegradation and photocatalysis studies of these polymers to understand their fate in marine environment and to develop eco-friendly green technologies to remove plastics litter [4, 5]. In this study, we investigated the behaviour and degradation of five biodegradable polymers (BPs): Poly(butylene succinate) (PBS); Poly(butylene succinate-co-adipate) (PBSA); Polycaprolactone (PCL); Polyhydroxy butyrate (PHB); Polylactic acid (PLA) in marine environment.

29.2 Experimental

29.2.1 Materials

Sample films of poly lactic acid (PLA), polycaprolactone (PCL), polyhydroxybutyrate (PHB), polybutylene succinate (PBS) and polybutylene succinate adipate (PBSA) were produced and placed in nets made of non-biodegradable material to perform the experiment. Mesocomes were used, which reproduce the natural conditions of the marine environment in the laboratory and three mesocosms are used for each polymer to create three replicates. Bacterial activators were also added to the mesocosms in order to reproduce an environment as close as possible to the natural marine environment.

29.2.2 Methods

The plastic films were placed in three different exposure conditions: (i) in the water column; (ii) placed on the bottom; (iii) buried into the sediment. The films were maintained for one year in the mesocosms and five film sampling sessions were carried out during the year to study the evolution of plastics in the marine environment. At each session, the weight of the films was recorded and a sample of the film was taken in order to see, firstly, the loss of mass during exposure, and secondly, to carry out analyses on the pieces of film.

29.2.3 *Analytical Technique*

The degradation of the selected polymers was monitored by gravimetric methods, Fourier transform infrared spectroscopy (FTIR), ^{1}H Nuclear Magnetic Resonance (NMR) spectroscopy, thermal analyses and scanning electron microscopy (SEM). The samples were analyzed on different days of exposure to different conditions.

29.3 Results and Discussion

29.3.1 *Results*

The polymers placed on the bottom, PLA, PBS and PBSA show a very low weight loss value, about 12.5%, 15.7% and 19.4%, respectively. PHB and PCL show higher values, about 92.5% and 85.6%, respectively (Fig. 29.1).

SEM micrographs of the samples after 363 days of degradation at different conditions (Fig. 29.2) were acquired in order to analyse the degradation phenomena on

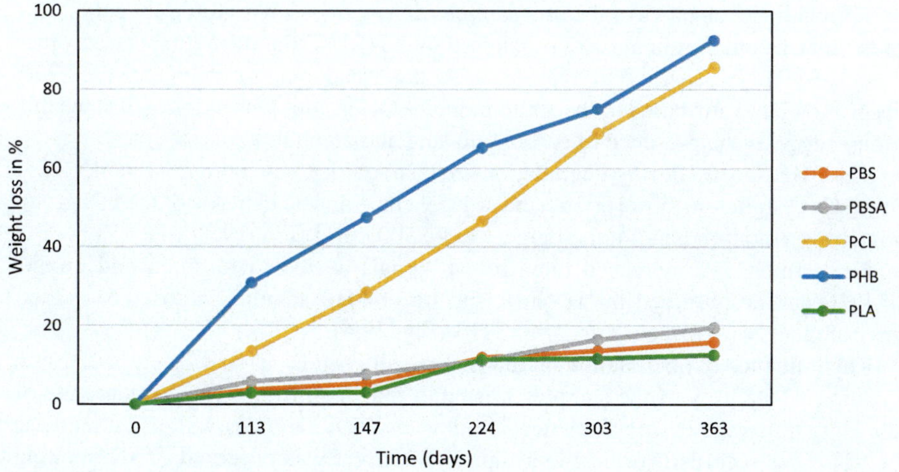

Fig. 29.1 Graph of the evolution of the mass loss (in %) over time of the polymers exposed on the bottom

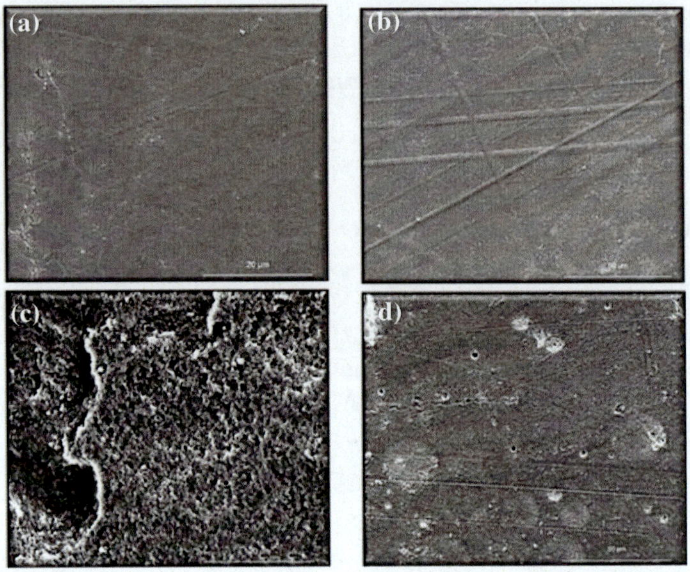

Fig. 29.2 SEM photography of PHB samples exposed in water column at t0 (**a**) and after 363 days (**c**) and PLA exposed on the bottom at t0 (**b**) and after 363 days (**d**)

the film surface. The PLA surface was slightly affected by exposure to the marine environment (Fig. 29.2b, d), whereas degradation phenomena were observed on the PHB surface (Fig. 29.2a, c). The formation of numerous heterogeneous pinholes, cracks and grooves was observed on the PHB surface suggesting the occurrence of degradation. PCL, PBS and PBSA are still being analyzed.

The chemical changes of the film samples during exposure to different conditions of the marine environment were monitored by FTIR-ATR and NMR spectroscopy.

Polyesters are sensitive to hydrolysis, so the broadening or change of the FTIR absorption band attributed to the stretching of −OH groups was monitored since this group suggests the presence of carboxylic acid and alcohol.

The PHB spectra during degradation are characterized by a progressive decrease of the C=O stretching absorption band at 1719 cm^{-1} and an increase of the OH group stretching vibration band in the range 3000–3500 cm^{-1} (Fig. 29.3).

Furthermore, an increased intensity of signals attributed to chain end groups of PHB can be observed in ^{1}H NMR spectra of PHB samples exposed to marine environment, with respect to pristine PHB (Fig. 29.4).

The influence of degradation on thermal transitions (T_g, T_c and T_m) as well as on the thermal stability of the samples was analyzed. A shift to higher temperature of the glass transition before and after 363 days of degradation as well as an increase in ΔH_m was recorded for all the samples. No effect was observed on the melting temperature.

All the samples present a similar thermal behavior during aging, and it is more evident for PHB samples (Fig. 29.5).

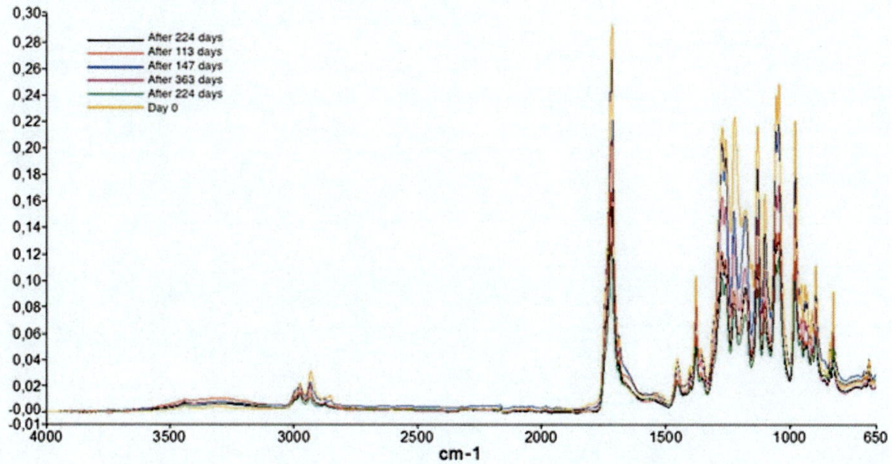

Fig. 29.3 Graph of FTIR-ATR spectroscope analyses of the evolution of the PHB film exposed in the water column

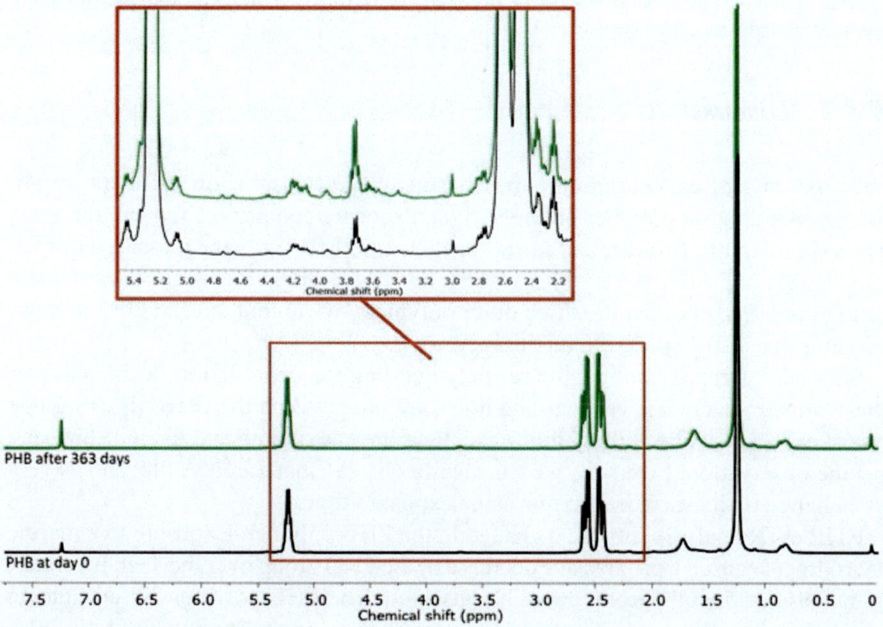

Fig. 29.4 ^{1}H solution NMR spectra (in $CDCl_3$) of PHB exposed in the water column at t0 and after 363 days

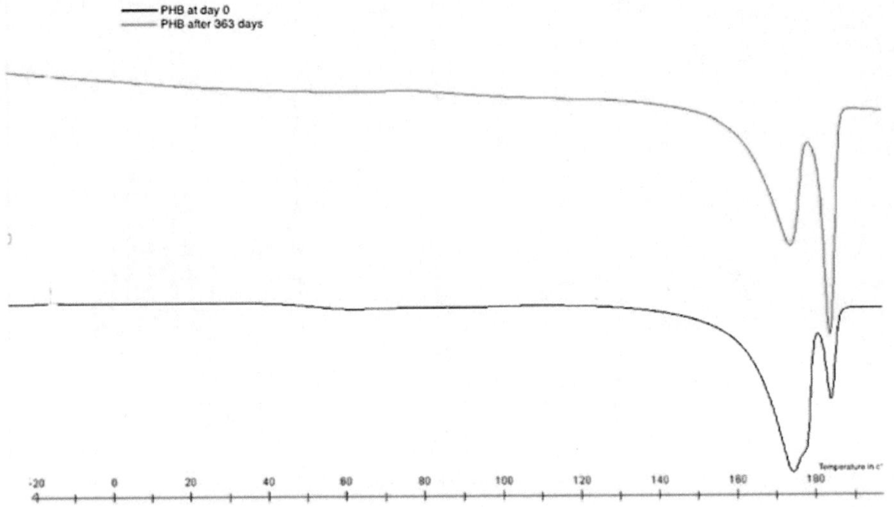

Fig. 29.5 Graph of Differential Scanning Calorimetry (DSC) analyses of PHB exposed in the water column at t0 and after 363 days

29.3.2 Discussion

After 363 days of exposure in the mesocosms, the polymers show different weight losses depending on polymer nature and the exposure conditions. Indeed, the mass loss is significantly stronger and faster for PCL and PHB while the mass loss rate for PBA, PBSA and PLA is very slow over time. The degradation is almost complete for PCL and PHB but for the three other polymers, we do not exceed 20% of mass loss after one year exposed in the mesocosms.

SEM photographs confirm the results regarding the degradation of the polymer films. Numerous cracks, grooves and holes are observed on the PHB film after one year of exposure while at the initial stage, the film was smooth and did not show any specific observation. Inversely, we can clearly observe that the PLA film shows very little change on its surface after the same exposure time.

FTIR-ATR analysis showed a change in the PHB polymer. Exposure to a marine environment caused a progressive decrease of the C=O group over the year. Inversely, an increase of the OH group could be observed. The PHB polymer did not remain insensitive to environmental exposure. These changes in the functional groups also induced the degradation of the polymer over time. NMR spectra further confirmed a change in signals attributable to terminal groups, probably produced by chain scission reactions through hydrolysis of ester groups.

29.4 Conclusions

The comparison reported in this work on the effects of exposure to 3 different conditions (in the water column, on the bottom and in the sediments) and the degradation of different biodegradable polymers leads to the conclusion that the biodegradable materials analysed are likely to degrade, under the experimental conditions used, at different rates. Although some biodegradable polymers (PLA, PBS and PBSA) show little degradation after one year of exposure, all sample analyses show degradation signals.

Acknowledgements The authors acknowledge Mrs. Maria Cristina Del Barone of LAMeST laboratory of the Institute for Polymers, Composites and Biomaterials, for her technical support in SEM analysis. This article is based upon work from the project - Biomonitoraggio di micro e nanoplastiche biodegradabili: dall'ambiente all'uomo in una prospettiva one health (BioPlast4SAFE) - Ministero della Salute.

References

1. Manfra L, Marengo V, Libralato G, Costantini M, De Falco F, Cocca M (2021) Biodegradable polymers: a real opportunity to solve marine plastic pollution? J Hazard Mater 416:125763. https://doi.org/10.1016/j.jhazmat.2021.125763
2. Wu WM, Yang J, Criddle CS (2017) Microplastics pollution and reduction strategies. Front Environ Sci Eng 11(6). https://doi.org/10.1007/s11783-017-0897-7
3. Viel T, Manfra L, Zupo V, Libralato G, Cocca M, Costantini M (2023) Biodegradation of plastics induced by marine organisms: future perspectives for bioremediation approaches. Polymers 15(12):2673. https://doi.org/10.3390/polym15122673
4. De Falco F, Avolio R, Errico ME, Di Pace E, Avella M, Cocca M, Gentile G (2021) Comparison of biodegradable polyesters degradation behavior in sand. J Hazard Mater 416:126231. https://doi.org/10.1016/j.jhazmat.2021.126231
5. Amato P, Muscetta M, Venezia V, Cocca M, Gentile G, Castaldo R, Marotta R, Vitiello G (2023) Eco-sustainable design of humic acids-doped ZnO nanoparticles for UVA/light photocatalytic degradation of LLDPE and PLA plastics. J Environ Chem Eng 11:109003 (2023). https://doi.org/10.1016/j.jece.2022.109003

Chapter 30
Resin Pellets Aging and Degradation Investigation from Long Term in Situ Experiment: First Results

Cristina De Monte, Marina Locritani, Silvia Merlino, Lucia Ricci, and Simona Bronco

Abstract Experimental studies in recent years highlight the presence of an increasingly high quantity of microplastics worldwide. The "resin pellets" represent a significant share among the first generation microplastics in the millimeter range (from 1 to 5 mm). They disperse in the environment, even unintentionally, during transport, storage and processing and recent studies show that their content varies from 3% to about 30% of all microplastics surveyed on beaches. A three-years experiment was carried out on a simulated beach and in marine water in Santa Teresa Bay (Gulf of La Spezia, Italy). In detail, special cages have been installed on the underwater observatory, LabMARE coastal station, placed at ten meters deep. The submarine station is equipped with a sensor for monitoring environmental parameters, recording data every 10 min. The experiment, aimed at investigating the behavior of plastic items and HDPE, PP, PLA and PBAT pellets, began on March 3, 2020. Here, the comparison between the properties of the raw pellets and those placed in the two different environments after six months, is discussed. Through chemical, spectroscopic and thermal analyses (GPC, SEM, FTIR-ATR, DSC, TGA) of granules, variations in color, surface morphology, chemical composition, thermal properties and molecular weight, and polydispersity of materials are analysed to show the different influences of environmental conditions.

C. De Monte (✉) · L. Ricci · S. Bronco
IPCF, Sede di Pisa del Consiglio Nazionale delle Ricerche, (CNR-IPCF), Via G. Moruzzi 1, 56124 Pisa, Italy
e-mail: cristina.demonte@pi.ipcf.cnr.it

M. Locritani
INGV, Via di Vigna Murata 605, 00143 Roma, Italy

S. Merlino
ISMAR, Sede di Lerici del Consiglio Nazionale delle Ricerche, (CNR-ISMAR), 19032 Lerici, Italy

M. Cocca et al. (eds.), *Proceedings of the 3rd International Conference on Microplastic Pollution in the Mediterranean Sea*, Springer Water,
https://doi.org/10.1007/978-3-031-34455-8_31

30.1 Introduction

Recent studies establish that plastic waste accounts for more than 75% of the total marine litter [1], and place the emphasis on the micrometric size fraction which is increasing in the coastal and in the marine environment debris findings [2–5].

However, in all the microplastics surveyed on the beaches, resin pellets represent a percentage ranging from 3% [6] up to 30% [7, 8] and are present in many seas.

Resin pellets are considered as the first generation microplastics and due to their loss during transport, storage and processing, they can enter the environment without limiting restrictions. Polyethylene (PE) and Polypropylene (PP) represent a large portion of plastic debris in the environment [9]. For this reason, the "International Pellet Watch" takes PE resin pellets as passive samples of POP pollution [10]. Many experiments and theoretical models are focused on the sorption of hydrophobic contaminants in microplastics collected in the marine environment [11–14].

Moreover, the use of biopolymers is increasing especially for commonly used single-use products and for packaging. European Bioplastic indicates that in 2021 bioplastics constitute the 1% of the 367 million tons of plastic produced globally, with a forecast growth trend of over 5 million tons in 2026 [15]. Among these, biodegradable plastics (including PLA, PHA, starch blends and others) account for over 64% of the global bioplastics production capacities in 2021, with 19.2% for polybutylene adipate-*co*-terephthalate (PBAT) and 18.9% for polylactic acid (PLA).

However, the objects and fragments stranded or recovered in the sea do not tell us the time they spent in the sea, their trip, their history before arriving there. These factors are important to evaluate the real effects due to the permanence of these materials in the marine environment even if it is not always easy to determine them. Therefore, most studies concerning microplastics aging and degradation are laboratory experiments carried out keeping the materials in artificial sea water [16] or in collected sea water [17, 18]. The simulated environment can only attempt to reproduce the physical and chemical characteristics of the real environment and lengthening the time of the experiment can help to minimize the differences in exposure to light and heat related to weather and seasonality.

Here, an experiment lasting three-years was carried out on a simulated beach (sandbox) and in a genuine marine environment in Santa Teresa Bay (Gulf of La Spezia, Italy) [19]. The experiment aimed at investigating the behaviour of plastic items and HDPE, PP, PLA and PBAT pellets in both environments. The experiment began on March 2020. The comparison between the properties of the raw pellets and those placed in the two different environments after six months is discussed in this work. The different influences of the environmental conditions are investigated by GPC, SEM, FTIR-ATR, DSC, TGA analyses of the granules, change in colour, surface morphology, chemical composition, thermal properties, molecular weight and polydispersity of the materials.

30.2 Experimental

Special cages (Fig. 30.1) were installed on the underwater observatory, LabMARE coastal station at 44° 4′ 55.08" N–9° 52′ 50.46" E placed at a depth of ten meters and about 60 m from the shore line [20]. The submarine station is equipped with a sensor for monitoring environmental parameters (SBE37SM, supplied by Sea-Bird Scientific), recording data every 10 min.

At the same time the same materials were placed in a sandbox (Fig. 30.2, located in Sarzana), containing sand taken from the Gulf of La Spezia, in order to carry out a comparative study on a "simulated beach".

Fig. 30.1 Particular of the cages installed on the underwater observatory before the deployment

Fig. 30.2 Particular of the sandbox

30.2.1 Materials

Four types of polymer pellets were used: high density polyethylene-HDPE, Auser Polimeri (Italy), polypropylene-PP, PoliEko (Slovenia), polylactic acid-PLA, Ingeo 2002D®, NatureWorks (US), and polybutylene adipate-*co*-terephthalate-PBAT, Ecoflex® F Blend C1200, BASF (Germany) and used as received.

30.2.2 Methods

30.2.2.1 Preparation of Samples

Pellets were put in two special stainless steel 318 cylindrical cages (about 40 cm × 30 cm) each containing three "baskets" (about 15 cm × 10 cm) installed on the structure of the underwater observatory (Fig. 30.1).

The sampling of the material under investigation was carried out pulling the station out of the sea. The materials were washed with seawater and with fresh water in laboratory before the analysis.

30.2.2.2 Analytical Techniques

Attenuated total reflectance (ATR) spectra were registered using a Fourier Transform—InfraRed Jasco 6200, equipped with a PIKE MIRacle accessory, 64 scans from 4000 to 650 cm^{-1} for each sample after the collection of background data. Thermogravimetric analysis (TGA) was performed with an SII TG/DTA 7200 EXSTAR Seiko analyser, under heating from 30 to 700 °C, at 10 °C/min rate and air flux at 200 mL/min, samples weight 5–10 mg. Differential scanning calorimetry (DSC) analysis was performed using a Seiko SII ExtarDSC7020 calorimeter with a different thermal programme for each type of polymer, samples weight 5–10 mg. GPC measurements were carried out on PLA e PBAT polymers by using an HPLC Agilent 1260 Infinity, with injection system at threeway valve and stationary phase at styrene–divinylbenzene resin, operating pressure of 80 bar, eluent $CHCl_3$ at 0.3 mL/min and detector at refractometer, and solutions of 3 mg/mL of samples in $CHCl_3$ (HPLC grade), two times filtered with an Agilent filter (2 μm porosity). SEM images were recorded with a FEI Quanta 450 ESEM FEG scanning electron microscope (CISUP Laboratories, University of Pisa). The specimens were analysed on the external surface without any specific manipulations.

30.3 Results and Discussion

30.3.1 Results

The analysis of the temperatures reached at sea and in the sandbox was considered an important factor to understand if their variability could influence the characteristics of the aged samples (Table 30.1) [19].

The collected data show a difference in the thermal excursion during the day between seawater and sandbox with a larger standard deviation calculated for sandbox temperatures data. This might suggest that materials in the sandbox could result much more thermally stressed.

Furthermore, the presence of algal and benthic fouling on the surface of the cages (Fig. 30.3) could have decreased the percentage of incident sunlight on samples in the sea and therefore have affected the oxidative degradation process, which is strongly influenced by the amount of incident ultraviolet radiation.

The visual observation of the pellets allows to notice a yellowing up to an amber colour for the samples in seawater and for PBAT also in the sandbox. SEM images show an increase in surface roughness.

Concerning thermal TGA analysis, the T_{onset} of the samples decreases after aging with more evidence for sandbox samples and for the PBAT (Table 30.2). No particular change in thermal transitions are observed by DSC analysis.

Table 30.1 Environmental parameters measured in LabMARE from march to august 2020 and in the sandbox located in Pisa from march to august 2021 [19]

	Average temperatures (°C)	Standard deviation[a] (°C)	T_{min} (°C)	T_{max} (°C)
Parameters measured in LabMARE				
March 2020	13.70	0.32	12.93	14.32
April 2020	15.07	0.78	13.69	17.33
May 2020	18.20	0.72	16.93	20.16
June 2020	20.88	0.85	18.97	23.19
July 2020	21.25	0.59	19.75	23.35
August 2020	23.68	2.20	22.12	25.24
Parameters measured in sand box				
March 2021	12.88	7.15	1.95	36.16
April 2021	14.98	6.45	0.123	36.21
May 2021	18.64	6.05	8.28	42.08
June 2021	25.64	7.48	11.79	45.60
July 2021	27.69	6.82	17.84	48.69
August 2021	29.06	7.19	18.81	51.63

[a]Calculated on all measures recorded in the month

Fig. 30.3 Surface of the cages after 6 months in sea

Table 30.2 T_{onset} of sandbox e sea samples from TGA analysis

	T_{onset} (°C)		
	Raw pellets	Pellet from sea	Pellets from Sandbox
HPDE	255.5	249.1	242.1
PP	261.5	261.4	255.0
PLA	323.1	320.2	315.0
PBAT	369.9	359.3	357.2

Absorption bands can be found in the ATR spectra in the region around 1100 cm^{-1}. In sandbox samples an increase in absorption bands due to double bonds and oxidation in the region 1700–1500 cm^{-1} is observed in all the cases with the exception of PLA. Peaks are also observed in the OH stretching region of both sea and sandbox samples.

The GPC analyses of PLA and PBAT allow to observe a strong reduction of the number average molecular weight for the PBAT sample sampled from the sandbox (from 21.5 KDa and 2.2 PDI to 12.6 KDa and 2.8 PDI).

30.3.2 Discussion

The increase in roughness, the changes in colour and the decrease in the thermal stability of the materials indicate the establishment of structural changes.

The appearance of the bands of the double bonds C=C and C=O can be related to phenomena of greater extent of photooxidation in the sandbox compared to the sea, while the appearance of the typical peaks due of –OH stretching can be referred to hydrolysis of the chains. The peaks at 1100 cm^{-1} are bands of contamination, due to residual sand.

The greater solar radiation and the higher temperature that characterize the environment of the sandbox are proved to be the factors with the greater influence on the aging of the materials than conditions in deep sea water.

30.4 Conclusions

The first results here reported are obtained over a period of six months from a long-term (3 years) in situ experiment on exposing microplastics (pre-production resin pellets) to sunlight and open sea and simulated sand conditions. The project aims to study the change of some of the chemical and physical characteristics of the structure of polymeric compounds during the aging in marine environments. Traditional plastics (HDPE and PP) and bioplastics (PLA and PBAT) are compared in a real, free and uncontrolled marine environment.

After 6 months of experiment (from March to September 2020) the materials immersed in the sea result to be subjected to less thermal stress than the corresponding ones in sandbox, because of the lower solar radiation and reduced thermal excursion.

Changes in the chemical-physical properties of the materials suggest a material aging and the beginning of a degradation process but with an evolution time that is still long and to be established.

References

1. Napper IE, Thompson RC (2020) Plastic debris in the marine environment: history and future challenges. Glob Chall 4(6):1900081
2. Peng L, Fu D, Qi H, Lan CQ, Yu H, Ge C (2020) Micro- and nano-plastics in marine environment: Source, distribution and threats—a review. Sci Total Environ 698:134254
3. Suaria G, Avio CG, Mineo A, Lattin GL, Magaldi MG, Belmonte G, Moore CJ, Regoli F, Aliani S (2016) The Mediterranean Plastic Soup: synthetic polymers in Mediterranean surface waters. Sci Rep 6:37551
4. Van Cauwenberghe L, Devriese L, Galgani F, Robbens J, Janssen CR (2015) Microplastics in sediments: a review of techniques, occurrence and effects. Mar Environ Res 111:5–17
5. Jung S, Cho S-H, Kim K-H, Kwon EE (2021) Progress in quantitative analysis of microplastics in the environment: a review. Chem Eng J 422:130154
6. Abelouah MR, Ben-Haddad M, Alla AA, Rangel-Buitrago N (2021) Marine litter in the central Atlantic coast of Morocco. Ocean Coast Manag 214:105940
7. Gündogdu S, Çevik C (2019) Mediterranean dirty edge: high level of meso and macroplastics pollution on the Turkish coast. Environ Pollut 255:113351
8. Merlino S, Locritani M, Bernardi G, Como C, Legnaioli S, Palleschi V, Abbate M (2020) Spatial and temporal distribution of chemically characterized microplastics within the protected area

of pelagos sanctuary (NW Mediterranean Sea): focus on natural and urban beaches. Water 12(12):3389
9. Teuten EL, Rowland SJ, Galloway TS, Thompson RC (2007) Potential for Plastics to Transport Hydrophobic Contaminants. Environ Sci Technol 41(22):7759–7764
10. Ogata Y, Takada H, Mizukawa K, Hirai H, Iwasa S, Endo S, Mato Y, Saha M, Okuda K, Nakashima A et al (2009) International pellet watch: global monitoring of persistent organic pollutants (POPs) in coastal waters. 1. Initial phase data on PCBs, DDTs, and HCHs. Mar Pollut Bull 58:1437–1446
11. Mato Y, Isobe T, Takada H, Kanehiro H, Ohtake C, Kaminuma T (2001) Plastic resin pellets as a transport medium for toxic chemicals in the marine environment. Environ Sci Technol 35(2):318–324
12. Endo S, Yuyama M, Takada H (2013) Desorption kinetics of hydrophobic organic contaminants from marine plastic pellets. Mar Pollut Bull 74(1):125–131
13. Karapanagioti HK, Endo S, Ogata Y, Takada H (2011) Diffuse pollution by persistent organic pollutants as measured in plastic pellets sampled from various beaches in Greece. Mar Pollut Bull 62:312–317
14. Rochman CM, Hoh E, Hentschel BT, Kaye S (2013) Long-term field measurement of sorption of organic contaminants to five types of plastic pellets: implications for plastic marine debris. Environ Sci Technol 47(3):1646–1654
15. European Bioplastics. https://www.european-bioplastics.org/market/ Accessed 14 July 2022.
16. Da Costa JP, Nunes AR, Santos PSM, Girão AV, Duarte AC, Rocha-Santos T (2018) Degradation of polyethylene microplastics in seawater: Insights into the environmental degradation of polymers. J Environ Sci Health 53(9):866–875
17. Brandon J, Goldstein M, Ohman MD (2016) Long-term aging and degradation of microplastic particles: Comparing in situ oceanic and experimental weathering patterns. Mar Pollut Bull 110(1):299–308
18. Bagheri AR, Laforsch C, Greiner A, Agarwal S (2017) Fate of so-called biodegradable polymers in seawater and freshwater. Glob Chall 1(4):1700048
19. De Monte C, Locritani M, Merlino S, Ricci L, Pistolesi A, Bronco S (2022) An in situ experiment to evaluate the aging and degradation phenomena induced by marine environment conditions on commercial plastic granules. Polymers 14:1111
20. DLTM. Alla-spezia-un-laboratorio-sottomarino-costiero-per-testare-nuove-tecnologie https://www.dltm.it/it/news/685-alla-spezia-un-laboratorio-sottomarino-costiero-per-testare-nuovetecnologie-4.html (2020). Accessed 06 July 2022.

Chapter 31
Quantification of Polyethylene in Mussel Hemolymph and Its Limited Additive Effect on Immune Function Induced by Bezafibrate

Concepción Martínez-Gómez, Marta Llorca, Tatiana Oporto, Silvia Rapuano, María del Mar García-Pimentel, and Marinella Farré

Abstract Bioaccumulation of waterborne pharmaceutical residuals may be influenced by the presence of microplastics in the surrounding water and therefore, this study investigated whether in vivo co-exposure to weathered polyethylene microplastics (PE) (nominal 1 mg/L) and bezafibrate (BZ) (500 ng/L) add-on effects on immune function in comparison to exposure to BZ alone, as it has been observed for other emergent contaminants. Mytilus galloprovincialis was used as model organism and the co-exposure study was carried out for 3 weeks. Pyrolysis–gas chromatography mass spectrometry (Py-GC/MS) technique was used for quantifying PE concentration in hemolymph. Lysosomal membrane stability (LMS) and phagocytosis efficiency of hemocytes, extracellular lysozyme activity and production of oxyradicals were determined as endpoints of immune function after 10 and 20 days. PE concentration in hemolymph of mussels treated with PE and PE+BZ were similar and had a high variability (hundreds of nanograms per milliliters) but values were one order of magnitude higher than mussels treated only with BZ, representing the first quantitative estimation of microplastic concentration that can be reached in this tissue under in vivo exposure conditions. However, still It remains to be determined what fraction of plastic particles were present in the plasma and what fraction carried by hemocytes. Mussels exposed to BZ and co-exposed to BZ+PE had significantly higher phagocytic rate than control mussels after 10 days, but this induction was transitory as phagocytosis efficiency was rather similar between treatments and control after 20 days. Lysosomal destabilization was found only after 20 days of exposure to BZ and BZ+PE, but not after exposure to PE. Overall, results obtained in this in vivo study

C. Martínez-Gómez (✉) · T. Oporto · S. Rapuano · M. del Mar García-Pimentel
Instituto Español de Oceanografía (IEO), CSIC, Oceanographic Centre of Murcia, Varadero 1, San Pedro del Pinatar, 30740 Murcia, Spain
e-mail: concepcion.martinez@ieo.csic.es

M. Llorca · M. Farré
Departament de Química Ambiental, Institut de Diagnosi Ambiental i Estudis de l'Aigua (IDAEA), CSIC, C/Jordi Girona, 18-26, 08034 Barcelona, Spain

M. Cocca et al. (eds.), *Proceedings of the 3rd International Conference on Microplastic Pollution in the Mediterranean Sea*, Springer Water,
https://doi.org/10.1007/978-3-031-34455-8_32

showed that the BZ+PE co-exposure did not exert synergic immunotoxic impact on mussels and that effects of BZ on immune function prevailed over the effects caused by the PE microparticles itself.

31.1 Introduction

The pharmaceuticals are introduced into surface waters and oceans due to incomplete removal of pharmaceuticals in wastewater treatment plants. Their occurrence in marine organisms raises concerns regarding toxic effects and development of drug-resistant genes [1, 2]. To date, pharmaceuticals detected in marine organisms and seafood include anti-inflammatories, antibiotics, oral contraceptives, and blood lipid regulators. Notably, drugs of the fibrate class are used as blood lipid regulators to treat hypertriglyceridemia. Fibrates act by stimulating nuclear receptors called peroxisome proliferator-activated receptors (PPARs). PPARs play an essential role in the general transcriptional control of numerous cellular processes, including lipid metabolism, glucose homeostasis, cell cycling, cell differentiation, inflammation, and extracellular matrix remodeling. Bezafibrate (BZ) is a lipid regulator that can be present as a contaminant in surface water in concentrations of ng/L [1, 3]. Some studies have revealed possible toxic effects in aquatic species caused by exposure to fibrates, ranging from general toxicity to endocrine disrupting effects and immune function [4–6].

The abundance of microplastics (MPs) in surface waters, sediments, and biota from the Mediterranean region are among the highest levels reported worldwide [7]. Therefore, bioaccumulation of waterborne pharmaceutical residuals can be influenced by the presence of microplastics in the surrounding water. The marine mussel *Mytilus sp.* presents some ecological and physiological characteristics that make it a good model for studying chemical contamination and contamination by microplastics, as well as an excellent model to investigate the effects of such contaminants on immune function [8]. The present study investigated whether in vivo co-exposure to weathered polyethylene microplastics and bezafibrate caused add-on effects on immune function in mussels (*Mytilus galloprovincialis*) exposed solely to bezafibrate, as it has been observed for other contaminants.

31.2 Experimental

31.2.1 Materials

Marine mussel *Mytilus galloprovincialis* [4–5 cm] were acquired from a Mediterranean myticulture farm located in Benalmádena (Málaga, Spain) (February 2021) and immediately transferred to the laboratory (Murcia, Spain), where they were acclimatized in a 500 L tank with aerated filtered natural seawater (FSW) (37 PSU and 19-20 ºC) until their use.

Polyethylene (PE) microparticles used in this study are produced for industrial purposes (commercially named MPP-635XF) and were acquired from Micro Powders, Inc. (USA). They have a crystalline (heterogeneous) form, a density of 0.96 g cm^3 at 25 ºC and mean size of 4–6 μm. MPs were suspended in filtered seawater (37 PSU) and stored in closed quartz glass Erlenmeyer flasks for four weeks. During this period, flasks were exposed to natural sunlight (July, 14 h sunlight/day) and outdoor ambient temperatures (ranging from 18 to 36 ºC). Afterward, the MPs were filtered, rinsed twice with FSW, and filtered to recover. Bezafibrate (CAS Number 41859-67-0) was purchased from Sigma-Aldrich. A stock solution of 101.8 mg L^{-1} in methanol was prepared and stored in the dark at 4 °C until use.

31.2.2 Methods

Five days before the experiments, mussels were translocated to 15L to continue the acclimatization period. The volume of FSW per mussel in the aquariums was adjusted to a ratio 1:4. Bivalves were exposed by triplicate for 10 and 21 days to four treatments:

- Control seawater (Filtered SW)
- PE = Polyethylene microparticles in FSW (1 mg L^{-1})
- BZ = Bezafibrate in FSW (500 ng L^{-1})
- PE + BZ = Polyethylene microparticles (1 mg L^{-1}) + Bezafibrate (500 ng L^{-1}) in FSW

In the case of PE + BZ treatment, PE microparticles were previously incubated with BZ in 200 mL of FSW using a glass bottle (24 h at 15 ºC) to allow potential adsorption of BZ on the PE plastic surface. After incubation, the suspension was added to the volume of FSW in the aquarium to get a final concentration of 1 mg L^{-1}. Aquarium water changes were performed 3 times a week, while PE load management and BZ dosage in the aquariums were performed every day and before feeding (*Tetraselmis suecica/Tisochrysis lutea;* adjusted to 1% of mussels dry weight). Bivalve mortality was checked daily. On days 0, 10, and 21, hemolymph samples were collected from mussels to measure immune parameters and concentrations of PE in the hemolymph.

31.2.2.1 Preparation of Samples

Hemolymph samples were obtained of a least 12 mussels for each treatment (four samples by triplicated). Hemolymph was withdrawn from the anterior adductor muscle. Individual primary cell culture samples (n = 12) were prepared for lysosomal membrane stability and phagocytosis analysis following the procedure described in [9]. Additionally, three hemolymph pooled samples (n = 4) by treatment were prepared to analyze extracellular lysozyme activity, extracellular reactive oxygen species (ROS) production, and PE concentration. Aliquots (250 μL) of pooled hemocyte suspension were diluted in an equal volume of TBS for ROS analysis. Aliquots of serum of hemocytes (centrifugation hemolymph pooled samples at 3000 g for 10 min at 4 °C) were used for lysozyme activity analysis. We measured PE concentration in pellets obtained from pooled hemocyte suspensions after centrifugation.

31.2.2.2 Analytical Techniques

Lysosomal membrane stability (LMS) was measured by the neutral red retention assay (NRR) following the procedure described by [9]. Lysosomal enzyme release by mussel hemocytes was evaluated by measuring lysozyme activity spectrophotometrically at 450 nM in the extracellular medium as described by [4]. The decrease in absorbance was continuously recorded at 450 nm for 5 min at room temperature. The results were expressed as Units of lysozyme ·mg protein^{-1}. Phagocytic efficiency (PhE) was measured using fluorescent microscopy (×400; Olympus BX43), following the methodology described by [9]. PhE was expressed as a percentage of cells that have internalized at least 3 fluorescent particles after counting at least 300 cells for each sample. The extracellular production of oxyradical by hemocytes was measured by the reduction of cytochrome c following the procedure described in [4]. Samples were read at 550 nm for 20 min using a TECAN SpectraFluor, and results were expressed as changes in optic density per mg protein.

Size exclusion liquid chromatography coupled to high-resolution mass spectrometry equipped with an Atmospheric pressure photoionization source (SEC-APPI-HRMS), working in negative conditions, was used to quantify PE concentration in the hemolymph. The separation of the polymer was based on toluene as a carrier liquid and separated in a polymeric size exclusion (APC) column. The acquisition was made in full scan mode working from 500 to 3,000 Da at a resolution of 17,500 FHWM. The quantification of PE was carried out by comparison with the calibration curve standard of a polymer of 1000 Da compared to samples. The quantification was based on equivalents of concentration.

31.3 Results and Discussion

31.3.1 Results

The control mussels had a low PE concentration at the start of the experiment (0.170 $\pm$ 0.087 ng mL^{-1}). However, control mussels, after 10 and 20 days, had values of PE concentration in hemolymph within the same order of magnitude as mussels exposed to 10 mg L^{-1} of PE (confidence interval for the mean 95% = 506.4–772.9 ng mL^{-1}). Low residual concentrations were found in mussels exposed only to BZ after 10 and 20 days (CI for the mean 95% = 13.1–69.0 ng mL^{-1}). Overall, a significant increase in PE concentrations over time exposure was not observed in any treatment except the Control SW (Fig. 31.1).

Moderate effects on lysosomal membrane stability were only observed in mussels exposed during 20 days to BZ and co-exposed to BZ + PE. The effects of co-exposure BZ + PE on LMS were more pronounced than the effects of exposure to PE, but significant differences were not proven (Fig. 31.2). Stimulation of phagocytosis was found in all treatments after 10 and 20 days of exposure, but it was significantly higher in mussels exposed to PE than in the rest of the treatments. Nevertheless, stimulation of phagocytosis was higher in mussels co-exposed to PE + BZ than in mussels exposed only to BZ (Fig. 31.1). In control samples, lysozyme activity in the extracellular medium ranged from 33 to 85 U μg prot.$^{-1}$. Overall, any treatment-related pattern was observed, although values close to zero activity were found after 20 days in mussels exposed to PE. Extracellular ROS production was similar in individual exposures (PE; BZ) and co-exposure conditions (PE + BZ) (Fig. 31.2). However, significantly higher ROS production was measured after 20 days than at time 0 in Control SW treatment.

31.3.2 Discussion

Interpretation of immune responses after exposure and co-exposure conditions of PE and BZ of this study are limited given that chemical analysis pointed to a PE contamination in control treatments along the experimental exposure, and a significant increase of phagocytosis activity and extracellular ROS production in Control SW treatment after 20 days. One possible explanation could be a human error caused by the dosage of PE in the control aquariums. Another potential explanation could be the aerial crossed-contamination between adjacent aquariums caused by handling during cleaning tasks, and by the aerosols produced during the air bubbling in the aquariums. However, low PE concentration found in control mussels at the beginning of the experiment and small residual concentration in mussels exposed only to BZ after 10 and 20 days give robustness to the quantification methodology and allow specific comparisons. This is the first quantitative estimation of microplastic concentration that can be reached in this tissue under in vivo exposure conditions.

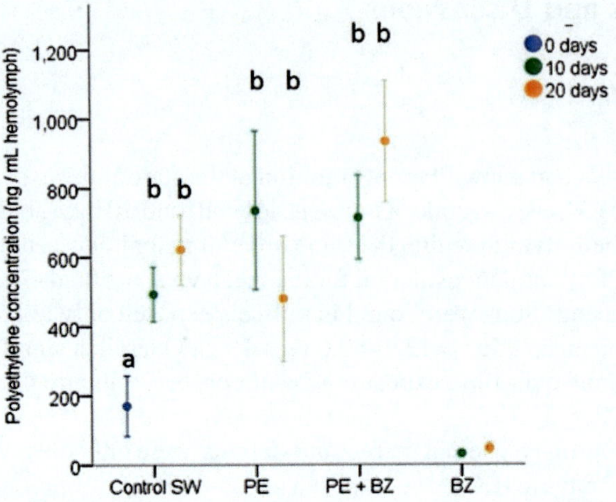

Fig. 31.1 Polyethylene concentration in mussel hemolymph (*Mytilus galloprovincialis*) exposed to control seawater (Control SW), waterborne polyethylene microparticles (PE) (1 mg L^{-1}, nominal concentration), PE and bezafibrate (BZ) (1 mg PE L^{-1} + 500 ng L^{-1} BZ) and BZ (500 ng L^{-1}) during 10 and 20 days

However, it remains to be determined what fraction of plastic particles was present in the plasma and what fraction carried by hemocytes.

Previous studies have proven that in vitro exposure to BZ stimulates phagocytosis in hemocytes of the freshwater mussel *Elliptio complanata* and marine mussel *Mytilus galloprovincialis* [10, 11], which is in agreement with our results. Furthermore, marine mussels in vivo injected with BZ showed a concentration-dependent lysosomal destabilization and extracellular lysozyme release [11]. In this sense, our results pointed to a cumulative effect of BZ on the lysosomal membrane over time, requiring an exposure of 500 ng/L for at least 20 days to observe moderate cellular stress on lysosomal destabilization (NRRT > 50 min). Co-exposure of mussels to PE + BZ stimulated phagocytosis response and decreased LMS in hemocytes to a greater extent than single exposure to BZ, suggesting some add-on effects on immune function, which were however not statistically proven. One possible explanation for the pattern observed could be non-linear bioaccumulation of BZ in mussels, as suggested by [11], and higher bioaccumulation of BZ in co-exposed mussels as a consequence of BZ sorption on PE particle's surface. In any case, chemical bioaccumulation data of this drug in mussels (currently ongoing process) are needed to fully elucidate the behaviour and kinetics of accumulation. However, longer exposure times would be necessary to fully understand if co-exposure to weathered polyethylene microplastics and bezafibrate caused add-on effects on immune function compared to exposure to bezafibrate alone.

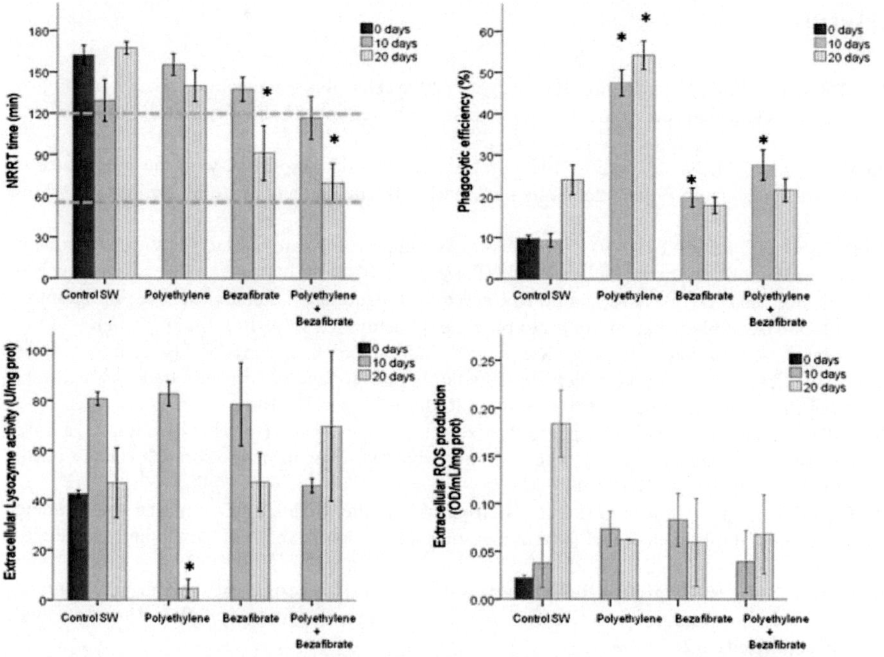

Fig. 31.2 Effect on immune function responses in mussel hemolymph (*Mytilus galloprovincialis*) exposed to control seawater (Control SW), waterborne polyethylene microparticles (PE) (1 mg L^{-1}, nominal concentration), PE, and bezafibrate (BZ) (1 mg PE L^{-1} + 500 ng L^{-1} BZ) and BZ (500 ng L^{-1}) during 10 and 20 days. Lysosomal membrane stability quantifies as neutral red retention time (NRRT). Data are the Mean ± Standard Error of three experiments (n = 4) by triplicate * = p-value ≤ 0.05 (significant difference against respective control). BAC = Background Assessment Criteria; EAC = Environmental Assessment Criteria

31.4 Conclusions

The occurrence of PE in mussel hemolymph is a direct consequence of its presence in the surrounding waters. Overall, this in vivo short-term exposure study showed that co-exposure to BZ + PE does not significantly alter the immune function of mussels in comparison to exposure to bezafibrate alone, and that the effects of BZ on immune function outweighed the effects induced by the PE microparticles themselves. However, results point that in vivo, at longer exposure times, the additive effects of BZ + PE on hemocyte parameters could be significant, and further research is necessary.

Acknowledgements The Spanish Inter-Ministerial Science and Technology Commission sponsored this study (PLAS-MED project; CICYT, CTM2017-89701-C3-3-R).

References

1. Metcalfe C, Miao X-S, Koenig BG, Struger J (2003) Distribution of acidic and neutral drugs in surface waters near sewage treatment plants in the lower Great Lakes Canada. Environ Toxicol Chem 22:2881–2889
2. Madikizela LM, Ncube S (2022) Health effects and risks associated with the occurrence of pharmaceuticals and their metabolites in marine organisms and seafood. Sci Total Environ 155780
3. Daughton C, Ternes T (1999) Pharmaceuticals and personal care products in the environment: agents of subtle change? Environ Health Persp 107:907–938
4. Canesi L, Ciacci C, Betti M, Fabbri R, Canonico B, Fantinati A, Pojana G (2008) Immunotoxicity of carbon black nanoparticles to blue mussel hemocytes. Environ Int 34(8):1114–1119
5. Rosal R, Rodea-Palomares I, Boltes K, Fernández-Piñas F, Leganés F, Gonzalo S, Petre A (2010) Ecotoxicity assessment of lipid regulators in water and biologically treated wastewater using three aquatic organisms. Environ Sci Pollut R 17(1):135–144
6. Velasco-Santamaría YM, Korsgaard B, Madsen SS, Bjerregaard P (2011) Bezafibrate, a lipid-lowering pharmaceutical, as a potential endocrine disruptor in male zebrafish (Danio rerio). Aquat Toxicol 105(1–2):107–118
7. Papadimitriu M, Allinson G (2022) Microplastics in the Mediterranean marine environment: a combined bibliometric and systematic analysis to identify current trends and challenges. Microplast Nanoplast 2(1):1–25
8. Canesi L, Ciacci C, Balbi T (2016) Invertebrate models for investigating the impact of nanomaterials on innate immunity: the example of the marine mussel *Mytilus spp.* Curr Bionanotechnol (Discontinued) 2(2):77–83
9. Martínez-Gómez et al (2019) In vitro effects of mercury (Hg) on the immune function of mediterranean mussel (*Mytilus galloprovincialis*) are enhanced in presence of microplastics in the extracellular medium. In: Cocca MC, Di Pace E, Errico ME, Gentile G, Montarsolo A, Mossotti R, Avella M (eds) Proceedings of the 2nd international conference on microplastic pollution in the Mediterranean sea, pp 27–33. Springer, Cham
10. Gagné F, Blaise C, Fournier M, Hansen PD (2006) Effects of selected pharmaceutical products on phagocytic activity in *Elliptio complanata* mussels. Comp Biochem Physiol C: Toxicol Pharmacol 143(2):179–186
11. Canesi L, Lorusso LC, Ciacci C, Betti M, Regoli F, Poiana G, Marcomini A (2007) Effects of blood lipid lowering pharmaceuticals (bezafibrate and gemfibrozil) on immune and digestive gland functions of the bivalve mollusc, *Mytilus galloprovincialis*. Chemosphere 69(6):994–1002

Chapter 32
Mechanical Recycling of Polyethylene-Rich Plastic Fraction Recovered from Marine Environment to Mitigate Plastic and Microplastic Pollution

Immacolata Liotta, Roberto Avolio, Rachele Castaldo, Giuseppe Andrea de Lucia, Andrea Camedda, Francesco Genovese, Gennaro Gentile, Gianluigi Tealdo, Maria Emanuela Errico, and Mariacristina Cocca

32.1 Introduction

Macro- meso- and microplastic pollution in the environment has become a global concern due to their impact on the marine ecosystem. Plastic production has increased substantially over the years and about 280 million tons are produced per year [1]. Plastics are widely used because of their properties and versatility, making them suitable materials for numerous applications [2]. Improper disposal of plastic waste has resulted in large quantities of these materials entering marine ecosystems [3] and the resulting impacts are becoming increasingly apparent. About 14 million tons of plastic end up in the ocean every year. Plastic waste is esteemed to represent 40–80% of the total amount of marine litter [4]. Polyethylene (PE) and polypropylene (PP) represent the most common polymeric fraction found in oceans [3]. When plastic ends up in the ocean, it breaks down into smaller and smaller fragments over time

I. Liotta (✉) · R. Avolio · R. Castaldo · G. Gentile · M. E. Errico · M. Cocca
Institute of Polymers, Composites and Biomaterials National Research Council of Italy, via Campi Flegrei, 34, 80078 Pozzuoli, NA, Italy
e-mail: immacolata.liotta@ipcb.cnr.it

G. A. de Lucia · A. Camedda
Institute for the Study of Anthropic Impacts and Sustainability in the Marine Environment, Località Sa Mardini, 09072 Torregrande, OR, Italy

F. Genovese · G. Tealdo
IREOS SpA, Via Stefano Turr 165, 16147 Genova, Italy

M. Cocca et al. (eds.), *Proceedings of the 3rd International Conference on Microplastic Pollution in the Mediterranean Sea*, Springer Water,
https://doi.org/10.1007/978-3-031-34455-8_33

due to environmental factors such as abrasion or exposure to sunlight, leading to the formation of microplastics: particles smaller than 5 mm [5, 6]. These particles appear to be ubiquitous, even in remote marine environments [7], and have significant impacts on ecosystems. For example, they can be ingested by marine organisms [8], hindering digestion and damaging the stomach [7]. They can also move from one trophic level to the next, leading to food chain contamination [9, 10]. Recovery and recycling of these materials play a fundamental role to mitigate marine plastic pollution and to remove from the environment a source of secondary microplastics. A rote to mitigate macro and microplastic pollution in marine environment consist into use, marine plastic debris as secondary raw materials to design recycled plastic materials. The mechanical recycling of marine plastic waste also poses a greater challenge than recycling land-based waste because marine waste contain a considerable amount of sand, salt, shells, algae and marine plants, that could hinder plastic recyclability [11]. In this work, polyethylene-rich plastic fraction (PE-rf) recovered from marine environment has been characterized through spectroscopic, thermal, and morphological analyses. A mechanical recycling approach was applied to realize recycled materials, using recycled polyethylene (R-PE) and different amount of PE-rf ranging from 100 to 40%. The effect of composition of recycled materials on the morphology, thermal and mechanical properties was analysed, demonstrating that the approach allows the valorization of the polyethylene-rich plastic fraction of marine litter.

32.2 Materials and Methods

32.2.1 Materials

A sample of PE-rf, recovered from marine environment, was kindly supplied by IREOS S.p.A. (Genova, Italy) and subjected to a sorting method. The material was mainly composed of small plastic pieces of various size and whole objects.

A recycled PE (R-PE) was used in this work to improve the mechanical properties of PE-rf.

32.2.2 Processing of PE-rf-Based Materials

PE-rf was subjected to cryogenic grinding in a Retsch mill (Retsch GmbH, Haan, Germany), using a sieve with 1 mm openings to obtain a fine powder. Ground PE-rf was melt-processed in a twin-screw extruder (Thermo Haake MiniLab) at 190 °C operating in continuous mode for 5 min. Firstly, a composition with 100% of PE-rf was made. A composition with 50% of PE-rf and 50% of R-PE was mixed by means of the extrusion process. In order to investigate the feasibility of a cold approach, another composition, with 40% of PE-rf and 60% of R-PE, was made by cold mixing.

After the mixing step, the materials were molded with compression moulding with 0.5 mm frame at 190 °C.

32.2.3 Techniques

Infrared spectra were recorded using a Spectrum 100 FTIR spectrometer (PerkinElmer, Waltham, MA, USA), equipped with an attenuated total reflectance accessory (ATR). The scanned wavenumber range was 4000–400 cm^{-1}. All spectra were recorded with a resolution of 4 cm^{-1}, and 4 scans were averaged for each sample.

Thermogravimetric analysis (TGA) was performed on a Pyris 1 TGA analyzer (PerkinElmer, Waltham, MA, USA) using air as purge gas and a linear heating ramp from 30 to 800 °C at 10 °C/min.

Differential scanning calorimetric analysis (DSC) was carried out on a TA-Q2000 system equipped with an RCS-90 cooling unit (TA Instruments, New Castle, DE, USA). Samples were heated from 20 to 200 °C; cooled from 200 to 20 °C; and heating from 20 to 200 °C using an heating/cooling rate of 10 °C/min.

Tensile tests were performed on dumb-bell specimens (0.5 mm thickness) at a cross-head speed of 5 mm/min by using an Instron 5564 testing machine (ITW Inc. Glenview, IL, USA). Young's modulus (E), peak stress (σ_{max}), and elongation at break (ε_R) were calculated as average values over at least 6 tested samples.

Scanning electron microscopy (SEM) was carried out on a Thermo Scientific Phenom Pure Desktop. Before SEM observations, cryofractured surfaces were sputter-coated with an Au/Pd alloy by means of an Emitech K575X sputtering device.

32.3 Results and Discussion

32.3.1 Preliminary Analysis of PE-rf Composition

Several PE-rf fragments were randomly selected and analyzed by means of infrared spectroscopy (FTIR). FTIR analysis was carried out according to the procedure described in Sect. 32.2.3. FTIR analysis qualitatively revealed the large presence of PE fragments and, in a lower amount, polypropylene (PP).

Melting temperature and thermal stability of the materials were determined using DSC and TGA.

Table 32.1 Results of mechanical testing: elastic modulus (E), tensile strength (σ_{max}), elongation at break (ε_R)

Sample	E (MPa)	σ_{max} (MPa)	ε_R (%)
PE-rf	1363 ± 51	11.03 ± 0.54	3.21 ± 0.08
PE-rf/R-PE 50/50	435 ± 69	11.50 ± 1.20	30.83 ± 23.19
PE-rf/R-PE 40/60	307 ± 21	7.59 ± 1.46	34.05 ± 13.80

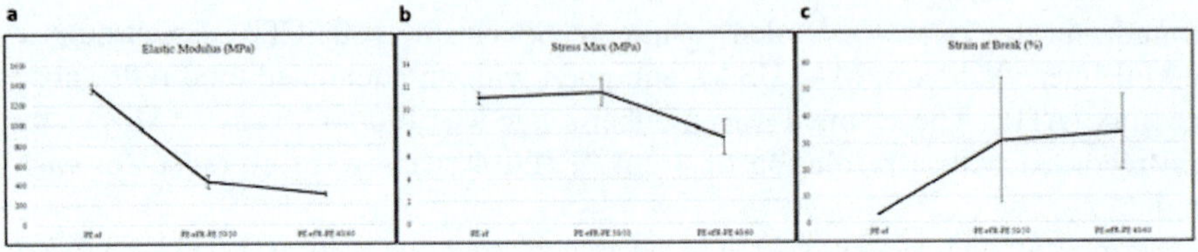

Fig. 32.1 Results of mechanical testing: **a** elastic modulus (E), **b** tensile strength (σ_{max}), **c** elongation at break (ε_R)

32.3.2 *Mechanical Testing*

Preliminary tensile tests of molded PE-rf showed (Table 32.1) a high elastic modulus compatible with the values usually exhibited by HDPE. However, high modulus is accompanied by low strength and very low elongation at break, around 3%. This behavior can be attributed to the presence of rigid impurities in PE-rf that hinder the deformation of the polymer matrix and lead to fragile behavior. The use of ductile R-PE as matrix has led to a significant increase in the strain at the expense of elastic modulus (Fig. 32.1). The effect on mechanical properties is due to the presence of chemically compatible filler which don't stiffen the composite.

32.3.3 *Morphological Analysis*

Morphological analysis of the fracture surfaces allowed to highlight the presence of inclusions in the morphology of the films produced (Fig. 32.2), due to non-homogeneity of PE-rf. The inclusions are however well dispersed in the matrix.

32.4 Conclusions

A recycling approach of a polyethylene rich fraction recovered from marine environment was presented in this work. The starting material was ground and several compositions with recycled polyethylene were prepared by extrusion process and compression moulding. Tensile tests results show that the use of a ductile polymer such as R-PE as matrix greatly influenced the strain at break at the expense of the

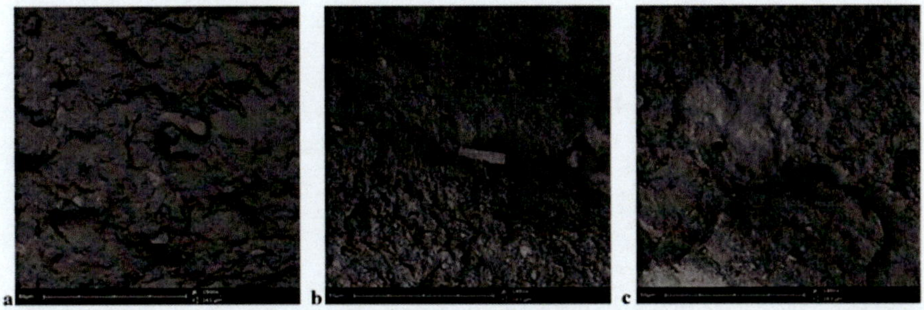

Fig. 32.2 Micrographs of cryofractured surfaces: **a** PE-rf, **b** PE-rf/R-PE 50/50, **c** PE-rf/R-PE 40/60

elastic modulus. Morphological analysis highlighted the presence of inclusions due to non-homogeneity of PE-rf which, however, are well dispersed in the matrix and don't cause stiffening of the composite. The effect of composition on the mechanical properties demonstrated that the approach allows the valorization of the polyethylene-rich plastic fraction of marine litter.

References

1. Adelodun AA (2021) Plastic recovery and utilization: from ocean pollution to green economy. Front Environ Sci 9(July):1–12. https://doi.org/10.3389/fenvs.2021.683403
2. Friot D, Boucher J (2017) Primary microplastics in the oceans | IUCN Library System
3. Prata JC et al (2019) Solutions and integrated strategies for the control and mitigation of plastic and microplastic pollution. Int J Environ Res Public Health 16(13):1–19. https://doi.org/10.3390/ijerph16132411
4. Almroth BC, Eggert H (2019) Marine plastic pollution: sources, impacts, and policy issues. Rev Environ Econ Policy 13(2):317–326. https://doi.org/10.1093/reep/rez012
5. IUCN, "Marine plastic pollution | IUCN. Issues Br., no. November, p. 1, 2021 [Online]. https://www.iucn.org/resources/issues-briefs/marine-plastic-pollution
6. Thompson RC et al (2004) Lost at sea: where is all the plastic? Science (80-.) 304(5672):838. https://doi.org/10.1126/science.1094559
7. Taylor ML, Gwinnett C, Robinson LF, Woodall LC (2016) Plastic microfibre ingestion by deep-sea organisms. Sci Rep 6(September):1–9. https://doi.org/10.1038/srep33997
8. Wright SL, Thompson RC, Galloway TS (2013) The physical impacts of microplastics on marine organisms: a review. Environ Pollut 178:483–492. https://doi.org/10.1016/j.envpol.2013.02.031
9. Mercogliano R, Avio CG, Regoli F, Anastasio A, Colavita G, Santonicola S (2020) Occurrence of microplastics in commercial seafood under the perspective of the human food chain. A review. J Agric Food Chem 68(19):5296–5301. https://doi.org/10.1021/acs.jafc.0c01209
10. Santonicola S, Volgare M, Di Pace E, Cocca M, Mercogliano R, Colavita G (2021) Occurrence of potential plastic microfibers in mussels and anchovies sold for human consumption: Preliminary results. Ital J Food Saf 10(4). https://doi.org/10.4081/IJFS.2021.9962
11. Avolio R et al (2019) Recycling polyethylene-rich plastic waste from landfill reclamation: toward an enhanced landfill-mining approach. Polymers (Basel) 11(2). https://doi.org/10.3390/polym11020208

Chapter 33
Does Calibration with Pristine PET Allow the Identification and Quantification of Aged PET Microplastic Particles by DSC?

Xhoen Gjashta, Benedikt C. Bulla, Sven Schirrmeister, and Kathrin Harre

33.1 Introduction

The microplastic (MP) pollution is rystalliz as one of the most challenging problems of the twenty-first century [1]. Due to inadequate waste management, plastic waste enters the natural environment, where it can remain for a long period of time [2]. During this time, the plastic waste is exposed to a variety of environmental stresses [3], which contribute to its degradation and thus to the formation of secondary MPs. Primary MPs originating from cosmetic beads or clothing fibers, enter the natural environment mostly through wastewater treatment. Once the MP particles enter water bodies, they move passively through rivers, seas, and oceans by complex physical flows [4]. Monitoring MP waste in water bodies and sediment can be challenging, due to the small MP particle sizes (<5 mm) [5]. Differential scanning calorimetry (DSC) is an analytical method for identifying and quantifying MP in environmental samples using the thermodynamic fingerprint of polymers [6–8]. A sample undergoes heating and cooling cycles at controlled rates, while its thermal rystall is monitored. The identification is based on matching of the thermal transitions of the respective polymer. The amount of an identified semicrystalline polymer in a sample can be calculated by dividing the value of the melting (or rystallization) enthalpy with the value of the specific melting (or rystallization) enthalpy of the polymer in the second heating (or cooling) cycle. Usually, pristine polymer material is used for the calibration to calculate the MP quantity [9]. MP detected in environmental samples has most likely been exposed to environmental stresses and has therefore aged naturally to some degree. The changes in material properties like

X. Gjashta · B. C. Bulla · S. Schirrmeister · K. Harre (✉)
University of Applied Sciences, Friedrich-List-Platz 1, 01069 Dresden, Germany
e-mail: kathrin.harre@htw-dresden.de

M. Cocca et al. (eds.), *Proceedings of the 3rd International Conference on Microplastic Pollution in the Mediterranean Sea*, Springer Water,
https://doi.org/10.1007/978-3-031-34455-8_34

chain conformation, molecular weight, molecular weight distribution or crystallinity can cause deviations between the calculated and the true values. In this study we analyse synthetically altered polyethylene terephthalate (PET) particles and fibers and compare their thermal properties with the pristine material. PET is one of the most common polymer wastes in the environment, as it is used to manufacture a range of short-lived consumer products from single-use bottles for beverages and other food products to synthetic fibers for clothing [10]. PET has heteroatoms in the main chain, allowing it to have higher thermal stability than polymers with a carbon backbone [10]. The most relevant pathways for PET degradation are photooxidative and hydrolytic degradation, which can lead to a reduction of molecular weight and an increase in carboxylic acid end group content. [11]. In this study, synthetically altered PET polymers were prepared by accelerated hydrolysis at 90 °C by heating a PET polymer suspension in aqueous NaOH. While accelerated photooxidation of PET under artificial UV irradiation in the laboratory is a widely accepted approach to produce synthetic altered polymers, accelerated hydrolysis can be a rapid and cost-effective degradation method to produce partially altered PET polymers relevant to various environmental studies [12]. The specific enthalpies of the partially altered PET polymers were compared with the pristine material to verify the acceptability of the calibration with pristine polymer material.

33.2 Experimental

33.2.1 Materials

PET fibers (PES hm 1,7/6) were purchased from STW (Schwarzwälder Textil-Werke) and recycled PET pallets from Kruschitz Plastics & Recycling. The PET particles were beforehand ground in a FRITSCH-cryogenic grinder (*analysette 3 SPARTAN pulverisette 0*), at 1–1,5 mm amplitude using a tungsten carbide grinding ball, while adding liquid nitrogen. The polydisperse grinded PET particles were subsequently dried at 60 °C until mass constancy. The particles were then sieved using a FRITSCH-sieve shaker (*analysette 3 SPARTAN pulverisette 0*) with an amplitude of 1 mm. The PET particle size fraction 100–200 μm was used for the experiments. Sodium was purchased from Merck Millipore.

33.2.2 Methods

In this study PET particles and PET fibers were also subjected to alkaline hydrolysis at different hydrolysis time. For each exposure, 200 mg of PET polymer were placed in a round bottomed flask containing 25 ml of aqueous NaOH (10 wt-%). The flask was equipped with a reflux condenser and preheated to 90 °C with an oil bath. The

Table 33.1 Overview of the experimental matrix

PET polymer	Sample	Hydrolysis duration in h	m_0 Amount of polymer pre-hydrolysis in mg	m_1 Amount of polymer after hydrolysis in mg	m_d Polymer Mass Reduction Rate in w%
PET fiber	F0,5	0,5	200,8	182,3	9,2
	F1	1	200,8	145	27,8
	F1,5	1,5	200,6	127,9	36,2
	F2,5	2,5	201	126,3	37,2
	F3	3	200,4	80,6	59,8
	F6	6	200,4	10,9	94,6
	F24	24	200	0,9	99,6
PET particle	P0,5	0,5	201,2	143,6	28,6
	P1	1	201,3	161,4	19,8
	P1,5	1,5	200	132,8	33,6
	P2,5	2,5	200,8	121,2	39,6
	P3	3	200,5	93,8	53,2
	P6	6	200,1	46,1	77,0
	P24	24	200,5	2,7	98,7

temperature was kept constant during the hydrolysis reaction. PET polymers were exposed to accelerated hydrolysis for 1; 1,5; 2,5; 3; 6 and 24 h. After the reaction, the flasks were transferred to an ice water bath to quench the hydrolysis reaction. A full overview of the experimental matrix is given in Table 33.1.

33.2.2.1 Preparation of Samples

After cooling, the suspension was centrifugated using a Hettich Centrifuge (*Universal 16A*) at 4000 rpm for five minutes. The aqueous supernatant was then decanted. The PET polymer was mixed with 25 ml of deionized water, the suspension was centrifuged again, and the aqueous supernatant was decanted afterwards. This process was repeated three times. The samples were then dried at 60 °C until a constant mass was reached. The partially altered PET polymer samples were stored in glass bottles until analysis.

33.2.2.2 Analytical Techniques

The thermal properties of the samples were haracterized using a NETZSCH DSC (*Polyma DSC 214*) under Argon atmosphere. The DSC instrument was calibrated for a temperature range of -50 °C to 300 °C with an indium standard. About 0,5 mg of the sample was placed in a closed aluminium DSC crucible positioned in the temperature-controlled DSC cell. Another empty crucible without sample was used as a reference. The sample was run through the calibrated temperature programme consisting of five different cycles [9]. The sample was first heated with 20 $Kmin^{-1}$ to 300 °C (I), followed by the isothermal stage (II), keeping a constant temperature of 300 °C for 3 min. The sample was then cooled with -20 $Kmin^{-1}$ down to -50 °C (III), followed by the second isothermal stage of 3 min (IV). Finally, the sample was heated again at 20 $Kmin^{-1}$ up to 300 °C (V). By monitoring the heat flow as a function of temperature, the phase transitions of the sample could be haracterized. The crystallization and melting peak were integrated using the NETZSCH Proteus Software (*Version 8.0.2*) to determine the specific enthalpies of the sample. The specific enthalpies are calculated as the quotient of the integrated enthalpy and the mass of the polymer sample weight into the DSC crucible. The 2nd heating curve was used to determine the specific melting enthalpy of the samples. An advanced method for determination of MP content uses the crystallization enthalpy and the melting enthalpy following a multiple regression [8]. The first heating and cooling cycles under controlled temperature rates allow the thermal history of the polymer to be erased. For each sample five different measurements were run on the DSC.

33.3 Results and Discussion

33.3.1 Results

Polymer Mass Reduction Rate. Six different samples of each PET fibers and PET particles were subjected to alkaline hydrolysis under the same conditions. The amounts of the altered PET polymer products were weighted, and the polymer mass reduction rates were calculated (Fig. 33.1). In the first three hours of hydrolysis, an almost linear correlation between the mass reduction rate and the hydrolysis time, is observed for both PET fibers and particles, which flattens out thereafter. 99,6% and 98,7% of PET fibers and particles respectively have been degraded after 24 h of hydrolysis. Due to the low amount of generated altered product, the 24 h hydrolysed PET polymer samples were not subjected to further analysis. Comparing these outcomes to the results of Sarno et al., a discrepancy between the values can be observed. Sarno et al. report a degradation of the PET fibers of 90% after 3 h of hydrolysis, as in our study a mass reduction of 90% is found after 6 h of hydrolysis. As the PET alkaline hydrolysis is strongly dependent from the polymer characteristics, these different results indicate that our PET fibers have a higher crystallinity, and that their specific surface area is lower.

Specific Melting and Crystallisation Enthalpies. The specific melting and crystallisation enthalpies were determined for each altered sample of both PET polymers

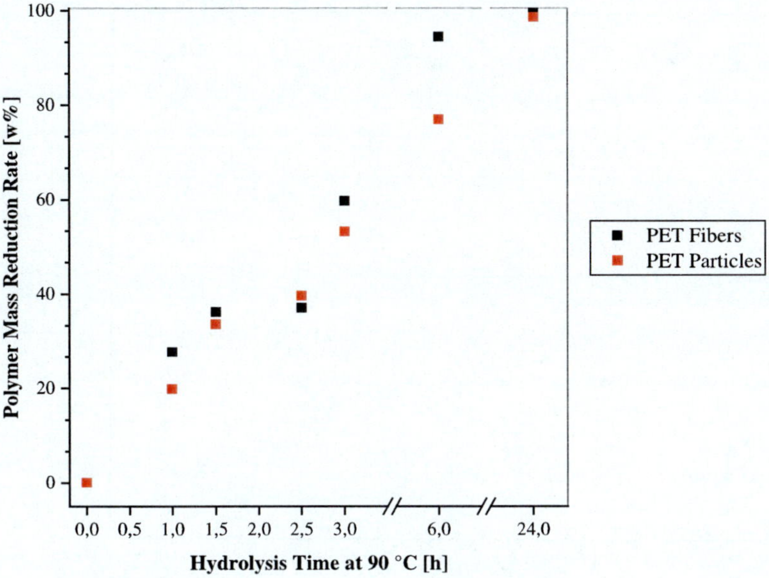

Fig. 33.1 The Polymer Mass Reduction Rate for PET Fiber (black) and PET Particle (red)

by a quintuple measurement. The values of the specific enthalpies are compared to those of the pristine material and are shown in Figs. 33.2, 33.3.

The specific enthalpies of most samples of the two PET polymers subjected to alkaline hydrolysis show no significant difference in their values, as they are all within the scatter range. Some single, samples, for example PET fibers partially degraded for 1,5 h until 3 h show specific melting enthalpies slightly higher than of pristine materials. To the contrary, the specific crystallization enthalpies for these samples, are slightly lower than the pristine material. By using the advanced multivariate regression model to calculate the MP content of a sample, the opposite discrepancies of the specific enthalpies can be eliminated [8]. These results indicate that the aged PET polymers can be characterised and quantified by DSC, which was calibrated with pristine PET polymer, without an underestimation of the MP particle content. The overestimation of the MP content will depend on the differences of the specific enthalpies, as far as indicated by our hydrolysis samples, as well as on the calibration method and the tolerance of the method.

From the values of the specific enthalpies, it can be deduced that the PET particles tested have a higher crystallinity than the PET fibers.

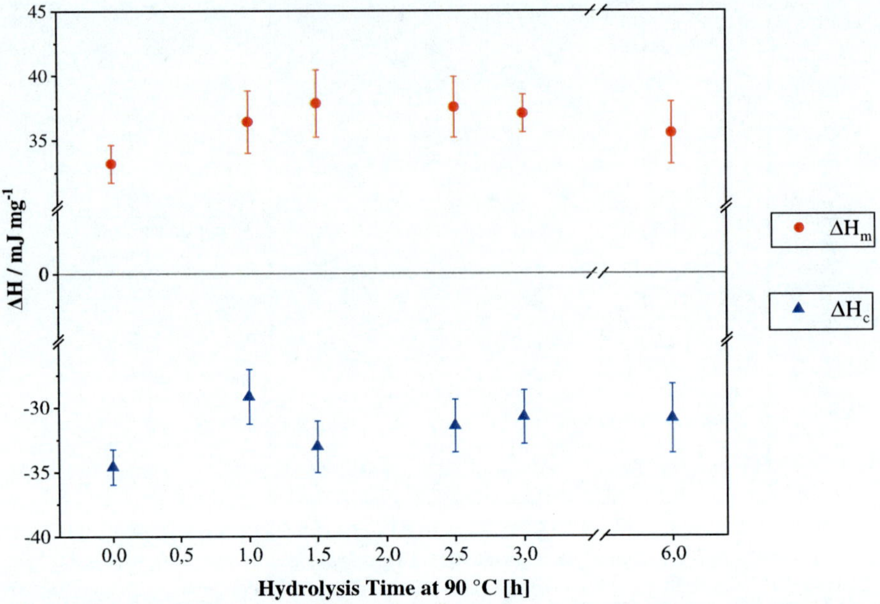

Fig. 33.2 Specific Melting Enthalpy (● ΔH_m) and Specific Crystallization Enthalpy (▲ ΔH_c) of pristine and partially degraded PET Fibers

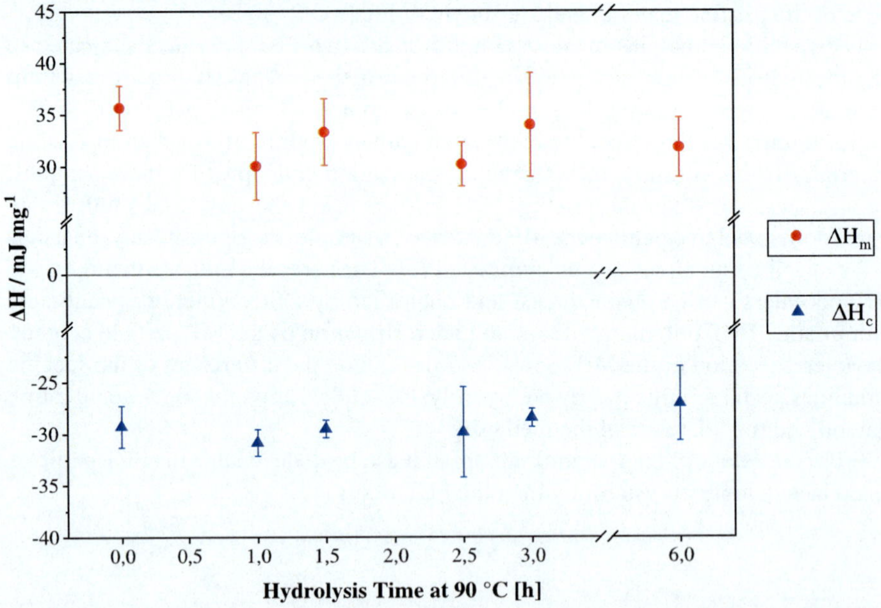

Fig. 33.3 Specific Melting Enthalpy (● ΔH_m) and Specific Crystallization Enthalpy (▲ ΔH_c) of pristine and partially degraded PET Particles

33.3.2 Discussion

The hydrolytic degradation of PET is the reverse reaction of the esterification of PET leading to the formation of its constitutive components terephthalic acid (TA) and ethylene glycol (EG) [11, 12]. An increase of hydroxyl and carboxyl end-groups in the polymer chains has been reported [10]. Hydrolytic degradation of PET is an autocatalytic degradation, when carboxyl end-groups are formed. [10]. Water is assumed to diffuse in the amorphous part of the polymer, which is strongly dependent on the morphology, crystallinity, and temperature [13]. A high degree of crystallinity hinders the reaction because the crystalline phase is inaccessible to water [10]. This is in accordance with the mass reduction rates, as the PET particles, which have a higher degree of crystallinity than the PET fibers, show a slower degradation rate. For example, 77% of the PET particles are degraded after 6 h of hydrolysis, while the mass reduction rate for the PET fibers, is at almost 95%.

Sarno et al. [12] report that even though the PET fibers were significantly modified physically after 3 h of alkaline hydrolysis, the FTIR spectra was almost identical to that of the pristine material. It has been suggested that the degraded products fragment from the main fiber and/or quickly dissolve in the surrounding aqueous media. A chemical degradation attacks the outer layer of the polymer particle [13], thus leaving the rest of the polymer matrix partially or incompletely degraded. Our results of the specific enthalpies for most of the aged PET polymer samples support these findings, as there was no significant difference in values from pristine material. A degradation of the amorphous part of the PET polymers by hydrolysis causes an increase of the crystallinity, thus allowing the values of the specific enthalpies to be higher. This is the case for the PET fibers subjected to hydrolysis for 1,5 h up to 3 h, which show significant higher specific melting enthalpies. As for the PET particles the insignificant differences of the specific enthalpies suggest that the hydrolysis was concentrated on the outer layer of particles, or it was inhibited from the characteristics of the polymer for example the lower specific surface area and/or the higher crystallinity. The reduction in mass is due to the formed degraded products, that might have dissolved in the surrounding aqueous media. As hydrolysis can lead to a reduction of the molecular weight, a shift in the melting temperature of the aged polymer would be expected [14, 15]. This thesis is not supported by the results of this study.

Sarno et al. report similarities between the physical and chemical properties of PET fibers degraded by alkaline hydrolysis and those exposed to UV light. However, a comparison to natural aged PET polymers is necessary, as it is suggested, that natural aging on MP can lead to other degradation states of polymers, since more complex degradation pathways take place [1]. The better understanding of natural aging and its effect on the thermal properties of polymers, is an objective of present research.

33.4 Conclusions

The results of this study suggest that the DSC method can be an effective analysing method to identify and quantify PET MP in environmental samples, if hydrolysis is the dominant altering mechanism. Partially degraded PET MP particles and fibers, that have been attacked and degraded mostly on their outer polymer layer can be analysed by DSC, which has been calibrated with pristine polymer material. But a general recommendation for a correlation between synthetic degradation and natural aging of PET polymers needs further investigation. The influence of different altering methods on the degree of degradation needs to be better understood. When analysing environmental samples for MP, the most likely occurring natural stress mechanisms should be taken into consideration. The authors recommend studying the influence of further altering methods on the thermal properties of PET.

References

1. Born MP, Brüll C (2022) From model to nature - A review on the transferability of marine (micro-) plastic fragmentation studies. Sci Total Environ 811:151389
2. Geyer R, Startz R (2017) Production, use, and fate of all plastics ever made, Science Advances
3. Sun Y, Yuan J, Zhou T, Zhao Y, Yu F, Ma J (1987) Laboratory simulation of microplastics weathering and its adsorption behaviors in an aqueous environment: A systematic review, Environmental pollution (Barking. Essex 265(2020):114864
4. Law KL, Thompson RC (2014) Microplastics in the seas, Science (New York, N.Y.) 345 144–145
5. Liu Z, Zhu Y, Lv S, Shi Y, Dong S, Yan D et al (2022) Quantifying the Dynamics of Polystyrene Microplastics UV-Aging Process. Environ Sci Technol Lett 9:50–56
6. Harzdorf J, Zeumer R, Schirrmeister S, Adomat Y, Kurzweg L, Faist S, Musche F, Grischek T, Harre K, Mikroplastik in sächsischen Gewässern, Schriftenreihe des Landesamtes für Umwelt und Geologie
7. Lucas Kurzweg, Sven Schirrmeister, Maurice Hauffe, Martin Socher, Kathrin Harre, (2022) Application of electrostatic separation and differential scanning calorimetry for microplastic analysis in particulate matrices, in: Setac
8. Sven Schirrmeister, Lucas Kurzweg, Martin Socher, Kathrin Harre, Identification and Quantification of Microplastics in Sediments using DSC, in: Micropol 2022.
9. Weise M, Großmann C, Bauer R, Grischek T, Socher M, Simon F, Harre K, (2021) Verfahren zur Quantifizierung von Polymerspezies in einer Polymerpartikel enthaltenden Probe(DE 1020200115971)
10. Venkatachalam S, SG, JV, PR, Rao, AK, Degradation and Recyclability of Poly (Ethylene Terephthalate).
11. Gewert B, Plassmann MM, MacLeod M (2015) Pathways for degradation of plastic polymers floating in the marine environment, Envitonmental Science Processes & Impacts
12. Sarno A, Olafsen K, Kubowicz S, Karimov F, Sait STL, Sørensen L et al (2021) Accelerated hydrolysis method for producing partially degraded polyester microplastic fiber reference materials. Environ Sci Technol Lett 8:250–255
13. Bacha A-U-R, Nabi I, Zhang L (2021) Mechanisms and the engineering approaches for the degradation of microplastics. ACS EST Eng. 1:1481–1501
14. Ronkay F, Molnár B, Nagy D, Szarka G, Iván B, Kristály et al. (2020) Melting temperature versus crystallinity: new way for identification and analysis of multiple endotherms of poly (ethylene terephthalate). J Polym Res 27
15. Horne FJ, Liggat JJ, MacDonald WA, Sankey SW (2020) Photo-oxidation of poly (ethylene terephthalate) films intended for photovoltaic backsheet. J Appl Polym Sci 137

Chapter 34
Mitigation Effect of Detergent on Microfiber Release

Nello Russo, Hector Alonso Fernandez, Maurizio Avella, and Mariacristina Cocca

Abstract In this study a mitigation approach based on the usage of a specifically formulated detergent is reported and a straightforward method for the evaluation of microfiber reduction is demonstrated. The effect of the liquid detergent in reducing the release of microfibers from synthetic fabrics during washing tests was evaluated.

34.1 Introduction

Microplastic released from textile fibers during the wash processes belong to emerging polluant among the different form of microplastics distributed in the environnement [1]. Several works investigated the role of additives (detergent, softener), washing temperature and washing time, different textiles parameters and the effects of water-volume on microfibres release [2–4]. Moreover, attempts have been made to reduce the detachment of microplastics during washing process using innovative finishing treatments [5–7] and to capture microplastics before their entrance in wastewater using innovative filtration systems [8–10].

In this study a mitigation approach based on the usage of a specifically formulated detergent with BASF ingredients is reported and a straightforward method for the evaluation of microfiber reduction is demonstrated. The effect of the liquid detergent in reducing the release of microfibers from synthetic fabrics during washing tests was evaluated. Washing tests of polyester fabrics were performed in absence/presence of two different detergents, a commercial product and a specifically formulated detergent, using low and high washing loads.

N. Russo · M. Avella · M. Cocca (✉)
Institute of Polymers, Composites and Biomaterials, National Council of Italy, Via Campi Flegrei 34, 80078 Pozzuoli, NA, Italy
e-mail: mariacristina.cocca@ipcb.cnr.it

H. A. Fernandez
Inditex - Industria de Diseño Textil, S.A, Avenida de la Diputación, Edificio Inditex, Arteixo (A Coruña), Spain

M. Cocca et al. (eds.), *Proceedings of the 3rd International Conference on Microplastic Pollution in the Mediterranean Sea*, Springer Water,
https://doi.org/10.1007/978-3-031-34455-8_35

Results indicate that the use of a specifically formulated detergent strongly reduces the amount of microfiber released from polyesters fabrics during both washings performed using low and high washing load. In detail, the amount of microfiber released during low washing load tests for taffeta pass from 0.269 ± 0.019 g/kg in the washing with commercial detergent to 0.040 ± 0.004 g/kg in the tests performed using the specifically formulated detergent. The amount of microfiber released in the washings performed using the specifically formulated detergent was close to that released during washing performed with only water confirming the mitigation effect induced by the usage of this specific liquid laundry detergent. The findings pave the way to a potential reduction of microplastic release as consequence of the usage of appropiate formulation for laundry product.

34.2 Experimental

34.2.1 Materials

For the washing tests two types of fabrics were supplied by Inditex in panels of about 89 × 75 cm (1 pcs): Taffetta with a continous and texturized warp yarn types and Jersey Black with a staple carded yarn (Fig. 34.1). Moreover, two kinds of detergent, one commercial (DET A) and the other one specifically formulated to limit the microplastics release (DET B), were provided by Inditex for the washing tests. Nylon filters were purchased from Merck Millipore.

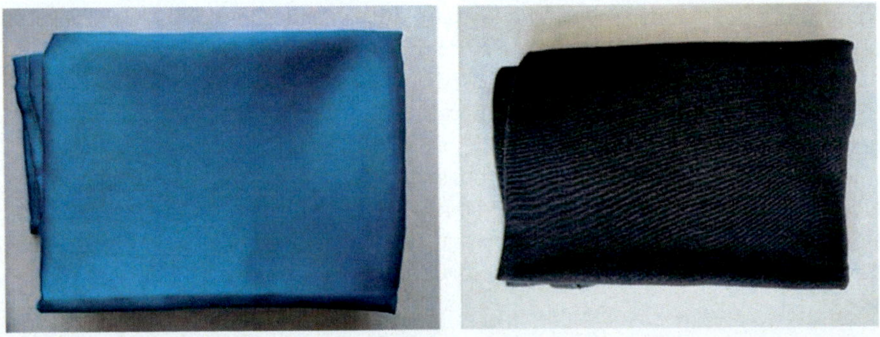

Fig. 34.1 Taffetta (left) and Jersey Black (right) supplied by Inditex

34.2.2 Methods

Washing tests were performed using a household washing machine Bosch Series 4 varioperfect WLG24225i, at 40 °C and 1000 rpm, the washing conditions of the program for mixed fabric using low (1 pcs) and high (1 kg) washing loads. An additional wash without fabrics was also run and the water was collected together with the previous washing water. Each test was carried out in triplicates (Fig. 34.2).

The water involved in each washing test was collected in tanks and filtered by means of a peristaltic pump (SP 311/60 Velp Scientifica) through nylon filters with 20 μm pore size. The amount of microplastics released during washing tests was gravimetrically determined. The filters used during filtering process were weighed before and after washing tests. Filter surfaces after filtration process were analysed under visible light microscope (Fig. 34.3).

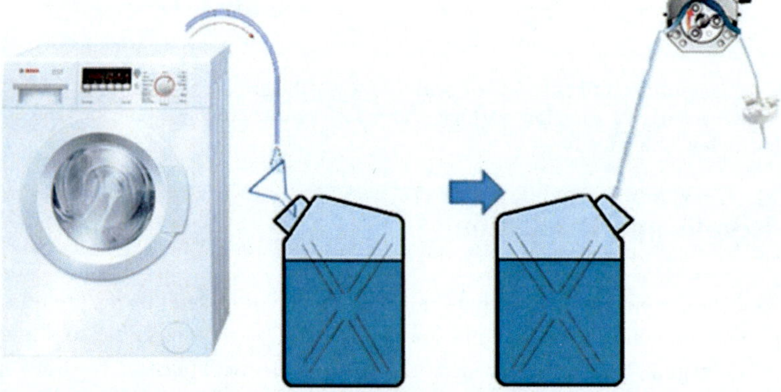

Fig. 34.2 Schematic overview of the used experimental apparatus

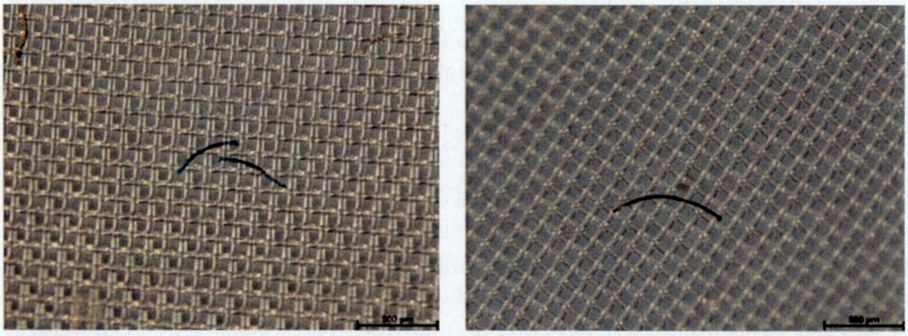

Fig. 34.3 Taffetta (left) and Jersey Black (right) microfibers seen under visible light microscope

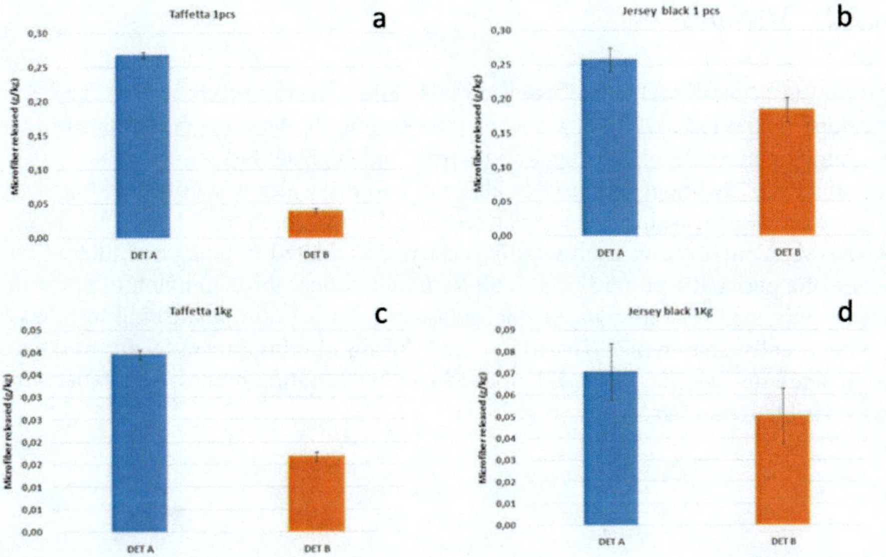

Fig. 34.4 Amount of microfibers released per kg of washed fabric: **a** Taffetta using low washing load; **b** Jersey Black (right) using low washing load; **c** Taffetta using high washing load and **d** Jersey Black using high washing load

34.3 Results and Discussion

In Fig. 34.4 the amount of microfibers released per kg of tested fabrics are reported. Looking at he bar charts for Taffetta and Jersey Black, it is possible to show that the use of detergent B strongly reduces the release of microplastics for both fabric types and used washing load. The reduction in the amount of released microfibers obtained in washing tests of Taffetta with detergent B was 85% in the case of low washing load (1 pcs), Fig. 34.4a, and 57% in the case of high washing load (1 kg), Fig. 34.4c. The reduction in the amount of released microfibers obtained in washing tests of Jersey black with detergent B was 28% in the case of low washing load (1 pcs), Fig. 34.4b, and 29% in the case of high washing load (1 kg), Fig. 34.4d.

34.4 Conclusions

Washing tests were performed on two kinds of fabric with two types of detergent and using two different washing loads. Detergent B resulted the most effective on microplastics release. This detergent represents the first product designed to reduce microfibers shedding from textiles.

Acknowledgements This work was supported by BASF for the development of the detergent formulation.

References:

1. Polluting Textiles The Problem with Microfibres Edited By Judith S. Weis, Francesca De Falco, Mariacristina Cocca. Routledge 2022
2. De Falco F, Gullo MP, Gentile G, Di Pace E, Cocca M, Gelabert L, Brouta-Agnésa M, Rovira A, Escudero R, Villalba R, Mossotti R, Montarsolo A, Gavignano S, Tonin C, Avella M (2018) Evaluation of microplastic release caused by textile washing processes of synthetic fabrics. Environ Pollut 236:916–925. https://doi.org/10.1016/j.envpol.2017.10.057
3. De Falco F, Pace E, Cocca M, Avella M (2019) The contribution of washing processes of synthetic clothes to microplastic pollution. Sci Rep 9:6633. https://doi.org/10.1038/s41598-019-43023-x
4. Volgare M, De Falco F, Avolio R et al (2021) Washing load influences the microplastic release from polyester fabrics by affecting wettability and mechanical stress. Sci Rep 11:19479. https://doi.org/10.1038/s41598-021-98836-6
5. De Falco F, Gentile G, Avolio R, Errico ME, Di Pace E, Ambrogi V, Avella M, Cocca M (2018) Pectin based finishing to mitigate the impact of microplastics released by polyamide fabrics. Carbohydr Polym 198:175–180. https://doi.org/10.1016/j.carbpol.2018.06.062
6. De Falco F, Cocca M, Guarino V, Gentile G, Ambrogi V, Ambrosio L, Avella M (2019) Novel finishing treatments of polyamide fabrics by electrofluidodynamic process to reduce microplastic release during washings. Polym Degrad Stab 165:110–116. https://doi.org/10.1016/j.polymdegradstab.2019.05.001
7. De Falco F, Guarino V, Gentile G, Cocca M, Ambrogi V, Ambrosio L, Avella M (2019) Design of functional textile coatings via non-conventional electrofluidodynamic processes. J Colloid Interface Sci 541:367–375. https://doi.org/10.1016/j.jcis.2019.01.086
8. De Falco F, Di Pace E, Avella M, Gentile G, Errico ME, Krzan A, ElKhiar H, Zupan M, Cocca M (2021) Development and performance evaluation of a filtration system for washing machines to reduce microfiber release in wastewater. Water Air Soil Pollut 232:406. https://biblioproxy.cnr.it:2481/10.1007/s11270-021-05342-6
9. McIlwraith HK, Lin J, Erdle LM, Mallos N, Diamond LM, Rochman CM (2019) Capturing microfibers—marketed technologies reduce microfiber emissions from washing machines. Mar Pollut Bull 139:40–45. https://doi.org/10.1016/j.marpolbul.2018.12.012.
10. Brodin M, Norin H, Hanning A-C, Persson C (2018) Filters for washing machines—mitigation of microplastic pollution. The Swedish Environmental Protection Agency. https://www.naturvardsverket.se/Documents/publ-kompl/1003-09-Report-Filters-for-washing-machines.pdf